Ideological Groups

Similarities of Structure and Organisation

R. KENNETH JONES
University of Botswana

Gower

Published by
Gower Publishing Company Limited, Gower House, Croft Road, Aldershot, Hampshire GU11 3HR, England

and

Gower Publishing Company, Old Post Road, Brookfield, Vermont 05036, U.S.A.

British Library Cataloguing in Publication Data

Jones, R. Kenneth
Ideological groups.
1. Knowledge, Sociology of 2. Ideology
I. Title
306'.42 BD175

ISBN 0-566-00634-0

Printed and bound in Great Britain by
Biddles Ltd, Guildford and King's Lynn

Contents

Foreword

There is a growing feeling among sociologists and others that some of the old divisions between traditional religions and secular movements are not appropriate. It is important for us to examine human beliefs and values as they actually operate, and without being too concerned with older definitional barriers. Consequently, it is refreshing that Dr Ken Jones, in looking at cosmologies, is concerned with the ways in which systems of meaning provide identity, and with ways in which secular movements can have, as it were, a religious significance. There is no doubt that at the human end it is the power of ideas rather than their origin which is important, especially in modern society when so many people are struggling to reaffirm their identity in a shifting environment. Dr Jones' study illustrates the connections between religion and other belief systems by an exploration of some very different groups ranging from the Socialist Party of Great Britain to the Swedenborgians. I think his material introduces us very well to a realistic appraisal in the way in which values have power and I would think that his approach is an excellent complement to recent studies such as those of Hans Mol and Bryan Wilson. It represents an insightful contribution to that more general field which may be known as worldview analysis.

1984

NINIAN SMART
University of California Santa Barbara
and University of Lancaster, England

Preface and Acknowledgements

For a number of years the author has been involved in the study of the sociology of religion, and especially the sociology of sectarianism. In particular he concentrated on a number of groups which, although ostensibly different in their ideologies, nevertheless appeared to resemble each other in a number of ways. This book gives an account of a number of such groups using original data based on field research.

An obvious danger with a study which attempts to encompass a number of different groups or categories is that it will become fragmented. The writer has endeavoured to avoid this and to keep within the limits of the work, namely to present a picture of some characteristics of organisation and structure of resocialisation groups and settings which are value-oriented, and to try to suggest reasons why such similarities exist in cosmological frameworks which are both secular and transcendental.

The work is intended primarily for sociologists of religion but it will certainly be of interest to students of small groups and in particular self-help groups.

Gratitude is expressed to a number of people who have given help, either formally or informally, in one way or another: Professor Anthony Coxon, Professor Michael Hill, Dr Roger O'Toole, Mr Peter Rolling (Shepherd), Professor Ninian Smart, Professor Roy Wallis, Professor Hans Mol, and Dr Bryan Wilson. It hardly needs to be stated that none of these bears any responsibility for the final form.

The writer would like to acknowledge his debt to the Editor of *Sociology* and the British Sociological Association, the Editor of the *Sociological Yearbook of Religion in Britain* and SCM Press, and to the Editor of *Sectarianism: Analyses of Religious and Non-Religious Sects* and Peter Owen Ltd, London, where sections of this work first saw the light of print. Appendix A was first presented at the World Congress of Sociology, Uppsala, 1978, and appears in a modified form in Mufti, R. and White, G., editors, *Understanding Social Change*, England: Nafferton Books (forthcoming), 1984.

Finally the author wishes to express his thanks to Athabasca University, Canada, which afforded the facilities for the volume's initial production by Marlene Paine, to Kathy Bowyer who typed it all, and to John Hobbs who helped enormously with the proof-reading.

Gaborone,
Botswana
1984

Introduction

There appear to be two broad areas in which groups which purport to control or change the identity of their members in some way resemble one another. These two broad areas of structure and organisation, together with the utilisation of a form of ideological legitimation, reveal broad similarities. Thus groups with principally a secular orientation often bear an uncanny resemblance in many respects to those groups whose stance is basically religious or transcendental. The task which we undertake here is to try to explore the nature of this similarity in six ostensibly very different groups.

A Study of Alcoholics Anonymous, the Catholic Apostolics, the Society of the New Church, Recovery, Inc., Neurotics Nomine, and the Socialist Party of Great Britain, revealed a considerable overlap between secular and transcendental value-oriented settings existing primarily as cosmologies which legitimated the lives of the members and as groups which managed the identity of members in some way. In particular such similarities were not surprising because, apart from the personal meaning systems of the individuals involved, members negotiate within the social process for the most part as members of collectivities or groups. Such groups constrain their members to enactments within certain areas and to particular recognisable patterns or regularities of behaviour. In other words, there exists a *minimum* of constancy of interactive patterns and symbol forms that are necessary for the group, whether concerned with a secular or transcendental ideology, to "put over" ideological legitimations, resocialisation procedures, and so on. Such groups have in common the minimum organisational requirements necessary.

Apart from similarities there were important differences in the groups in terms of specific social processes which served to distinguish them and mark off those which were designated transcendental from those overtly secular. Moreover, there were broad differences to be observed in each group's origins, patterns of development or growth, degree of separation from the world, proselytisation, and so on.

Through the years, a man peoples a space with images of provinces, kingdoms, mountains, bays, ships, islands, fishes, rooms, tools, stars, horses and people. Shortly before his death he discovers that the patient labyrinth of lines traces the image of his own face.

(J.L. Borges, 1972, *A Personal Anthology*, London: Pan Books)

"In any man who dies there dies with him
His first snow and kiss and fight.
It goes with him.

There are left books and bridges
And painted canvas and machinery.

Whose fate is to survive.
But what has gone is also not nothing:

By the rule of the game something has gone.
Not people die but worlds die in them."

(Yevtushenko)

1 General problems of orientation

Comparatively recent work in the sociology of religion has appeared to suggest that there is too sharp a distinction drawn between transcendental and non-transcendental belief systems, although there does emerge what seems indeed to be quite clearly a continuum of belief systems from the purely empirical to the completely metaphysical. Moreover, this continuum is one which appears in all respects to be a two directional one. Processes which are in essence religious or transcendental, or are normally associated with these, often emerge in secular or non-transcendental processes.[1] Similarly, secular or non-transcendental processes may and often do appear in overtly transcendental processes. It may be, of course, that the similarity of organisation and structure may be due to the process of secularisation which would ultimately result in religious and non-religious processes being indistinguishable from one another. On the other hand, the similarity of structure and organisation are due to the fact that empirical systems, including belief systems, correspond, i.e., actual group structures and organisations overlap. Perhaps also, in studying transcendental organisations, sociologists should not expect to find variables and processes so very different from other important areas of human activity. It may well be the case that human processes are historically and socially attributed different "labels" although in essence they are, or are almost, identical. Certainly, to some extent, religion is now regarded not as a uniform phenomenon but as a recurring pattern, and the concepts which are employed to categorise the elements of this pattern rather than locating elements of this pattern in the universe have simply a heuristic usefulness.

In a sense, it may be true to say that what comprises religious behaviour is that which comprises non-religious behaviour; in other words the differentiation lies not in the constituents of the empirical realm, but in the frame of reference or perspective through which we view the empirical realm. The religious congregation is not essentially different from other congregations; the church building is not essentially different from other buildings; the objects of symbolisation are not essentially different from other objects; the religious sect not essentially different from other sects. There is nothing which is itself specifically religious or transcendental; religion "claims the whole world of experience" as do equally science and art.[2]

Religion or transcendentalism legitimises the universe of the observer. The most important task of the sociology of religion "is to analyse the cognitive and normative apparatus by which a socially constituted universe is legitimated".[3] The sociology of religion is thus forced to consider other legitimating systems which are religious or pseudo-religious.

The observer differentiates human behaviour and social groups into categories such as transcendental and secular, and often this differentiation is both useful and valid. But in terms of sociological analysis it may well be the case that apparently dissimilar movements may possess similar systems of organisation and structure. Etzioni's normative model of organisations such as educational institutions, voluntary associations, and religious organisations - organisations where normative power was the main agency of control - is not fine enough. There is a complicated belief structure which relates to the normative system which involves goals and means which can relate equally to both transcendental and non-transcendental normative systems.

The group under study, both transcendental and non-transcendental, are ordered systems in society governed by or possessing discernible rules or relationships of systems. The structure and organisation of such groups are assumed to be segmented and ordered. This does not, of course, imply that "system" in the social situation mirrors or parellels an "order" in "nature", nor a Levi-Straussian search for "universals", but merely that there is a constancy of interactive patterns and symbol forms that are necessary for the existence and functioning of groups. Both the transcendental and non-transcendental groups under study possess a cognitive and normative apparatus which legitimate the universes of these groups. The groups are legitimating systems for the actor or observer whether they are transcendentally orientated or not.

The social "forces" or agencies which give rise to these groups - loss, abandonment, stigma, isolation, deprivation, and so on - generate primary ("religious") ideas, sentiments, management techniques, structure and so on. This is not to suggest that there exists an "underlying principle of human association"[4], but rather something more akin to the interacting of relatively "fixed" symbol systems. The similarity of a transcendental group to a non-transcendental group in terms of organisation, patterns of interaction, etc., is a factor measured by the frequency in which the groups share the same or similar characteristics as measured by agreed criteria. Symbolic systems of interaction sustain and perpetuate an ordered system of offices through which the role incumbents pass, which is what we mean by the social structure of groups. Groups are dispositionally geared to stable interaction processes involving specific normative expectations. The size and propinquity of membership of groups are also determining factors in the systematic and cultural properties of a group situation.

Problems of terminology

Weber has suggested that it is not possible at the start of a presentation to define religion. "Definition can be attempted, if at all, only at the conclusion of the study". Furthermore, Weber even goes so far as to suggest that we ought not to be concerned with the "essence" of religion, merely the "conditions and effects of a particular type of social behaviour". The variability of the many kinds of religious behaviour leads us, he suggests, to examine the subjective elements of such behaviour in terms of "meaning".[5]

Firstly, however, it appears obvious that some idea of what we imply by "religious behaviour" ought to be entertained before we embark upon a study. We can even go so far as to say that such an idea is imperative and indeed necessary - that we begin to sort out and distinguish the phenomena of the external world. Secondly, not only is it doubtful that the notion of "essence" needs to be contained in a definition of religion but such a quest can lead the searcher into deep and problematic water - even into metaphysics. Thirdly, the suggestion he makes that religious behaviour can best be examined in terms of "meaning" or the "grounds of meaning" which shape our social action is an extremely fruitful one, because it enables us to focus not only on the possibility of arriving at a sociological definition of religion but also on the interpretive stances adopted by social participants.

Is a definition of religion possible?

The total range of phenomena considered as religious by the host of sociologists and anthropologists is so vast and varied that we doubt whether any simple statement or attempt at a definition can cover them all.

Thus Martin [6] says that "if there are no exclusive associations between one polar alternative and any related set of alternatives, no sets of criteria can be utilised to distinguish between the religious and the secular. There would be an obvious absurdity in combining the metaphysical and mythopoeic modes of thought, acceptance of miracle, belief in historical purpose, rejection of material benefits, and lack of confidence towards the world, under the common rubric of religion". In fact, as we now well know, he obviates the distinction between "this worldliness" and "other worldliness". An additional argument is used by him to try to show that from a wide range of twenty alternatives, only four (transcendental, other worldly, next-worldly and supernaturalistic) are non-secular [7] although we might argue that these also would depend on the meaning which we attributed to them. What Martin is saying, basically, is that certain identities of structure are common to both "secular" and "religious" or "transcendental" concepts.

Some modern theologians, notably Tillich and the Christian Atheists, have argued that a postulation of a religious area of experience in opposition to a secular one is an indication of the basic separation of man's spiritual life from his depth. Several have argued for the elimination of the substantive term "religion" to cover, or rather dichotomise, these.

Bultmann talks of religion as a form of human longing to escape from the viscissitudes of this present world to another world which can accommodate the soul, whereas Gerhard Ebeling talks of an "enlargement of reality by means of God". More recently a sociological and linguistic sophistication has made many contemporary theologians wary of a broad definition.

In terms of sociology, the very diversity of the subject areas covered bring into question the nature of the phenomena of "religion". For example, recent studies have dealt with the nature and substance of belief as well as unbelief, with political groups, with social movements such as the John Birch Society, with the nature of secular and religious ritual, with scientific and religious belief systems, and so forth.

Attempts at some functional definitions of "religion" perhaps in terms of its integrative power, no longer yield much in the way of fruitful analysis, partly because of the very pluralism which is characteristic of modern industrial societies. Instead, many have moved towards asking such questions as "How is religion used?", or "How does it fit into the stream of social life of which it appears to be a part?" The abandonment of the quest for substantive reality has led some to examine in detail the possibility of an inherent structural reality, embedded in all men, which seeks religious expression, while others tend to view religion as a "Cultural reality" in no sense "given" but as a "creatively expressed" reality manifested in symbolic manifestation.[8]

The quest for a general definition of "religion" to encompass phenomena as diverse as the belief systems of the Nupe, the UFO'ists, or the complexity and diversity of many social movements, assumes a fixed and identifiable perspective of this particular phenomena. On the contrary, we are concerned not with generic abstractions but symbolic representations linked to institutional and group settings. "Each 'religion' as the symbolic (and cultural-institutional) vehicle for maintaining a faith is specific to the particular group - tribe, community, society - which shares it."[9]

If we return to Weber's discussion of religion, we see a marked lack of enthusiasm in dealing with all orientations of culture in his definition, and therefore he tended to exercise an exclusivist (narrow) definition which claimed the identification of certain characteristic phenomena as religious. In contrast to this type of definition Durkheim set the field for a broad and inclusive (and functionalist) categorisation. It therefore becomes permissable, in the Durkeimian tradition, to treat ostensibly non-religious social phenomena such as humanism, psychoanalysis, and other 'isms' as well as movements in which members exhibit a high degree of total commitment, as religious phenomena of the same kind. We deal elsewhere with the claim that such phenomena are in fact either functionally equivalent to religion or religious surrogates.

Many social scientists now subscribe to an inclusivist definition. Geertz [10], for example, defines religion as "a system of symbols which acts to establish powerful, pervasive and long-lasting moods and motivations in men by formulating conceptions of a general order of existence and clothing these conceptions with such an aura of factuality that moods and motivations seem uniquely realistic". Spiro [11] talks of it as "an institution consisting of culturally defined patterned interaction with culturally postulated super-human beings" and makes the interesting point that "while religious beliefs are not always of ultimate concern, non-religious beliefs sometimes are". Thus, if "sacred" refers to objects and beliefs of ultimate concern, then "religious" and "secular beliefs alike may have reference either to sacred or profane phenomena". He quotes the members of Kiryat Yeddim, and Israeli Kibbutz, which have all the characteristics of a religious belief system with the exception of super-human beliefs. The same applies to other manifestations of ultimate concern, such as communism, baseball and so on. He goes on to add that "They may, of course, serve many of the functions served by religious beliefs; and they are, therefore, members of the same functional class. Since, however, they are substantially dissimilar, it would be misleading to designate them by the same term." Spiro's definition is obviously much more exclusivist than Geertz's despite the fact that his illustrations "let in" a certain inclusivist element.

Bellah [12] defines religion as "a set of symbolic forms and acts which relate men to the ultimate condition of his existence". Such a definition derives its nature from the work of Paul Tillich who offers a generalised and compromising stance in relation to the question of "ultimacy". But besides this interesting theological heritage and legitimation such a definition is essentially a very limited functionalist approach. The same kind of approach can be seen in others, such as Herberg [13], who talks of religion as "the structure of one's being orientated to one's ultimate concern".

The early development of an approach that was later to become both more sophisticated and pervasive in the hands of others is contained in the work of John McMurray [14]. He argues that religious experience begins in human experience, and "the special kind of behaviour that we call religious is composed of the same elements as is behaviour which is not religious, its only distinguishing characteristic being that it offers a field of experience about the world which is partial, temporary and dependent. The centre of the field of religious experience is the experience we have of others in relation to ourselves". He goes as far as to say that what is reflected in our religion "is the general forms and attitudes of our personal relations with one another". This appears to be uncannily echoed in the work of Peter Berger and Thomas Luckmann. For Luckmann in particular, everything human is also at the same time religious, because by religion he means the capacity of the human organism to transcend its biological nature through the construction of universes of meaning which are morally-binding, objective and all-embracing. In a recent essay, he has argued that all socialisation is essentially religious, but some is more religious than others. This is a version of the functionalist approach because religion is defined in terms of its function - what it actually does for the individual or society. For Luckmann, it is unfortunate (although an historical fact) that certain patterns of behaviour are commonly recognised as specifically religious, but this is merely because "Definition and interpretation of religious representations is held in monopoly by religious institutions which may also enforce 'correct' religious performance".[15] Berger, although similar to Luckmann in the identification of the sociology of religion with the sociology of knowledge, posits a sacred cosmos (a universal and meaningful order of things). Berger argues that the human situation contains "signals of transcendence" which he defines as "phenomena that are to be found within the domain of our 'natural' reality but that appear to point beyond that reality...(the transcendental) is the transcending of the normal, everyday world".[16] Earlier, in conjunction with Luckmann, he had argued that Language has a transcending power - "the transcendences have spatial, temporal and social dimensions," and earlier still they had both argued for the identification of the sociologies of knowledge and religion, and that the religious system was but one legitimating system.[17]

Berger's approach is useful for this particular study because he highlights the human tendency to impose upon a meaningless universe a meaningful order of reality. This tendency to "get right with the cosmos" legitimates the stance of human beings and adds an additional "supra-human" endorsement. In a pluralistic society, there can be a plurality of cosmos, which can be secular as well as transcendent. He further draws attention to the nature of religion as being a particular kind of ideology.

In terms of this study the definition (or rejection of the possibility of a meaningful definition) will play a large part in determining the path taken and the answers we try to arrive at.

Religious boundaries

We feel that in view of the diversity of definitions and areas of study covered by sociologists in their quest of the study of religion, it is difficult to arrive at a satisfactory general definition of religion - or even the desirability of this. Furthermore, just as there is an increasing tendence to obviate the distinction between magic and religion in primitive societies, and the apparent lack of an internal distinction between the supernatural (super-empirical) and the natural (empirical), so it becomes clear that it is not so easy, in these societies anyway, to draw precise and clearly demarcated boundaries between the areas of human behaviour we might call religious and those areas we agree are profane. Evans-Pritchard,[18] for example, criticises Durkheim's distinction between the sacred and the profane on the grounds that the two categories "are so closely intermingled as to be inseparable" rendering the dichotomy a useless one. Goody is concerned with this, and the larger question, of what it is which is involved in "categorising acts and beliefs as religious". Part of the answer is seen by Goody as establishing that the criteria for "the isolation of sacred or ritual or magico-religious phenomena are derived not from the actors but from the observer's assessment of what is intrinsic". Both Goody and Horton [19] arrive at a definition of religion as entailing a belief in supernatural beings, and such a variable is also supported by Spiro who argues that it is the "one variable that ought to be designated by any definition of religion". However, belief in the existence of super-human beings does not, according to him, necessarily entail that religious behaviour has to be otherworldly or "spiritual", for the simple reason that such superhuman beings can be means or ends in the total scheme. We have already quoted with approval his claim that religious beliefs are not always of ultimate concern. He goes on to further state that it is not the case that religion, and religion alone, refers to the "sacred" while the secular must always concern itself with the realm of the profane for "religious and secular beliefs alike may have reference either to sacred or profane phenomena".

The work of Mandelbaum [20] distinguishes between the transcendental and pragmatic functions of religion, and he claims that in India villagers generally segregate these categories quite sharply, delegating different deities, rites and practitioners to each. Thus the two complexes of religious beliefs and practices are taken as complementary to each other. Religion becomes, for Mandelbaum, "all that is included in a people's relation to the supernatural, as they conceive the supernatural".

The work of Spiro, Goody, Horton and Mandelbaum casts considerable light on the whole boundaries or territories of demarcation between sacred/secular, pragmatic/transcendental, this worldly/other worldly, and so on. Perhaps the important contribution is to talk of the problem in terms of actor and observer, that is, an actor frame of reference and an observer frame of reference. Thus what to an actor may appear to warrant the designation "religion" to the observer might very well not do so. Various rituals, beliefs and meaning systems in our society which the actors or participants may not dream of calling "religion" may, to the observer, be very worthy of the designation. The idea of actor and observer is not a new one. Nadel used it very effectively in his study of Nupe religion, particularly in his argument that the "category of the supernatural is of dubious value in anthropological description".[21] But when we talk or impute communication or control of the supernatural to others, "we are clearly viewing the situation through the eyes of the people we are studying, imputing to the thought of transcending the given reality; we should probably impute to them also definite attitudes and feelings relating to the things supernatural, such as awe or fear, humility or ecstasy, or the emotional concentration we call worship Yet when we try to define the supernatural itself, we are driven to assume the opposite viewpoint and to judge the transcendentality of things by our own way of thinking".[22] Furthermore, "the very conflict between supernatural and empirical knowledge on which we base our judgements are again valid only within our own system of thought". The point at issue is that in our (the observer's) society access to two sorts of credibility (faith _and_ knowledge) is available. The actor, however, is not facing a choice "between faith and knowledge, having to set aside one when he accepts the other. Rather he accommodates both, only distributing them over different provinces of life".

Nadel's discussion of the "competences" of religion, later expanded upon by Worsley,[23] demonstrates quite clearly that such "competences" (to avoid the trap of falling into thinking of "functional requisite"), or at least the first three (experiential explanation of the universe; economic ethics; support of the social structure) can be replaced by secular agencies. It is doubtful, too, whether religion alone can provide individuals with specific experiences and stimulations, which is the fourth competance. This area, in which Nadel speaks of the _especially_ religious, is itself open to grave doubt. As Worsley comments, such especially religious experience is not only discouraged in some cultures, but very few experience it anyway, even in cultures where it is socially approved. "It is, then, by no means a universal attribute of religion, and cannot be an irreducible part of the analytic definition of religion." Further Worsley tries to kill the category, or at least its special duality, altogether, by the death of a thousand qualifications; "But as the other three elements are only competential, and not necessarily found, what does religion consist of as distinct from other kinds of behaviour?" The answer seems to lie for Worsley with Spiro's definition that "what distinguishes religious belief from other kinds of social belief is that, in some way or other (and that way may be oblique, remote or tenuous), it refers to, and looks for validation in, a dimension beyond the empirical-technical realm of action".

It seems, then, that so far we are faced with two main areas dealing with the nature of religion as an analytical category, the one being _inclusivist_ or broad enough to allow in forms of individual and collective behaviour which might not immediately appear to be religious, and the other being _exclusivist_ or narrow enough to posit a number of seemingly inflexible criteria which help us to order the phenomenon...which we identify as "religious". The latter, or exclusivist, mode of analysis appears to place a great deal of emphasis on the nature of the "supernatural", "transcendental", "other-worldly" and so on, and not only does it place such an emphasis but it both claims to recognise such phenomena and even purports to be able to define it. A second problem is the variance in the available evidence as to whether simple societies do distinguish as readily as anthropologists make out between "supernatural" and "secular" realms, or whether in fact it is much more difficult to draw such boundaries than we might at first suppose. After all, even in the modern industrial society such distinctions are by no means easy to make. In all this discussion we fall into a way of talking about the religion of a society or group which after all there may be very little to connect Azande with Nupe religion, and even less with the formalised ritual of the Anglican Church. It may be the case that there is more in common between the May Day ceremonies of the Russian State and the Anglicans than between many forms of religion in primitive societies, both in the areas of belief and of practice.

Substantive and functional definitions of religion

Peter Berger has argued recently [24] that the transcendent, because it has to be observed, examined, verified and so on empirically, must of necessity appear as immanence, echoing an old theme of Sapir and Whorf (and later Wittgenstein) that the limits of our senses are the limits of our world. The dimension of dreams is not accessible through empirical means. Human meanings may possess qualities other than human projections onto the transcendental.

The two major definitions have fallen into those that regard religion as substantive (the content of meaning which forms the phenomenon) and those that regard it as functional, (what religion does in the social or psychological system). The British empirical tradition in the social sciences has relied heavily on the functionalist approach of Durkheim and Malinowski. Three sophisticated modern theories are represented in the work of Bellah, Geertz and Luckmann which share in common, according to Berger, a way of looking at religion as "defined in terms of what it does".

The concern of Berger is that such definitions "slice up reality" into what should be included in a definition of religion and what should be excluded. "Thus substantive definitions of religion generally include only such meanings and meaning-complexes as refer to transcendent entities in the conventional sense - God, gods, supernatural human beings and worlds...By contrast, functional definitions are likely to include such meaning-complexes as nationalism or revolutionary faiths with the result that the commanalities...of the contemporary counterculture are brought into focus - and, by the same token, the differences between them are relegated to inattention."[25]

The importance of the functional definition lies in its utility - "for example, investigations of the socio-psychological mechanisms by which this or that worldview is maintained as plausible in the minds of its adherents". Thus, in the present study, the emphasis is essentially on a broad functional definition of religion to include <u>other than</u> transcendent entities, partly in order to examine the plausability-maintenance systems, the management of identity, and the coherent formulation of alternative world views or cosmologies.

Berger, however, sees the shift towards functional approaches as a "quasiscientific legitimation of the avoidance of transcendence in which the special nature of religious phenomena is avoided by equating it with other phenomena. The religious phenomena is 'flattened out'. Finally, it is no longer perceived. Religion is absorbed into a night in which all cats are grey". This is Berger's prerogative. It may well be the case that such a position highlights the inescapable fact that such a "flattening out" has the inevitability which emerges from a new cosmic comprehension - that is the way in which the vast majority have come to see the world. All experience is valid, and on the same incontrovertible existential plain. Also, it may well be the case that secular manifestations of transcendental characteristics are legitimate examples of the "domestication" of ecstacies, chanelling them "into socially acceptable and useful activity". In such a manner we arrive at the management of identity in cosmological settings.

Religious surrogates

So far we have delineated two major approaches to the definition of religion, the nominal-inclusivist-functional, and the real-exclusivist-substantive, in the attempt to show not that our secular phenomena is to be equated with religious phenomena because neither possess the quality of the supernatural but that such a quality of the supernatural, <u>as we understand it</u>, spills over in abundance from within the confines of traditional locations and is to be found instead in many common areas of human activity. This is not a particularly new approach for it can be traced even as far back as Simmel who is the antecedent of a more contemporary rendering found in Berger, Goffman and of course Luckmann.

In <u>Die Religion</u>, Simmel sets the scene as follows: "The experiences man encounters in a certain mood stir relations, meanings, sentiments, of which themselves are not yet religion, nor do their realities in any way conform to the religion of a differently attenuated soul; but divested of this reality and forming in themselves a sphere of objectivity, they become "religion" which here means the "objectified world of faith".[26] Religion becomes, for Simmel, something created or constructed by individuals out of non-religious feelings and experiences. This perspective is very different from the concept of <u>religionswissenshat</u> where religion is imminently experienced and felt. Rather, it is in the perspective of <u>religionssociologie</u> in which alternative aspects of reality directly express religion.

The unity of perception with which we view society means that Simmel is unable to agree to the possibility of experience of religion or meaning <u>away from society</u>. To Simmel religion is merely another form of social interaction and is to be studied wherever it is to be found, and in whatever form it manifests itself. In two very important points he appears to sum up our discussion so far. Firstly, religious feelings, impulses, emotion and so on, are not found exclusively in religious systems: there is no monopoly of any brand or variation of feelings; secondly, it is an impossible task to try to search for the distinguishing, basic, or other, characteristics of religion. It is in the sexual act and at the grocer's just as much as the crypt and the village church.

Berger, as we mentioned above, has argued that we "rupture the fabric" of ordinary life by alighting upon "signals of transcendence" which lead us to an additional dimension away from this ordinary life. Thus honour, music, sexuality, pure mathematics, the propensity for order and so forth, are designated as signals of this sort. Similarly, Goffman's emphasis on ritual and face-work is suggestive of an additional dimension. The savings of the social face is a dramaturgical task.[27]

Luckmann's view is that the very transformation of our biological nature is a religious phenomenon.[28] This democratisation of religion has its roots in the very formation of the self, in the socialisation process, which echoes Simmel's words that God's unity "documents itself in symbol and approximation in the organism and the social group".

The question of religious surrogates is a problematic area because by designating something as a surrogate, we are assuming we know that it is a surrogate <u>of</u>. Further, we suggest that it functions or surrogates in approximately <u>the same</u> way as that of which it is a surrogate.[29] We deal elsewhere with the thorny problem of function, need and the "fixed human-nature" perspective which this implies. Suffice it to say that at this juncture, very rarely does a functional surrogate fulfill the role of functional equivalence. One of the important contributions which Durkeim makes is that religious supernaturalism is socially constructed rather than <u>sui generis</u>. "The world of religious things is not one particular aspect of empirical nature; it is superimposed upon it." The supernatural becomes, therefore, something not normally found in the natural world but something which our attention is drawn to because of its essentially bizarre nature. Perhaps this can be better phrased as something which imbues ordinary interaction with qualities (order, coherence, depth, meaning, totality - or anything else) not normally found there, as we generally view it. The supernatural becomes, therefore, the transfiguration and symbolic expression of society.

We are now in a position to suggest some tentative criteria of religion. Firstly, there seems to be a capability of transcendence by transfiguration and symbolic expression. Secondly, religion or religion-like phenomena, postulate a cosmic unity or world story. Thirdly, there seems to be a possession of revealed knowledge of a special kind. By means of one or more of these criteria definitions become <u>shared</u> definitions of the situation whereby the sacralisation process of giving permanence, stability and magnitude to established patterns of behaviour becomes in order. The mechanisms for the sacralisation process can be roughly identified as follows: Mol [30] talks about a) objectification, in which we give coherence and reference to variegated elements of our existence by locating a transcendental reference point which confers order, consistence and timelessness; b) commitment, in the sense of an emotional "latching on" to a specific aspect of identity; c) ritualisation, which produces social cohesion and a focus for the renewal of sentiment; d) myth, or the depicting of man's symbolic nature in a short-hand account. In more general terms, some secular movements approximate to the status of religions because they share common characteristics in organisation and structure and do the <u>same sorts of things</u> for their adherents. Usually, they lay claim to knowledge of a special kind based on authority which goes beyond the individual member. In addition, such movements harness the diversity of individual members and fashion such diversity into radical coherence by means of the introduction of a normative order which members feel obliged to follow. In other words, such movements expressly manage the identity of members in significantly similar ways, usually in their claim to primary authority in their revelation of a world-view. Lastly, such groups often offer sanctuaries for the disorientated individual who comprise them, acting as powerful symbolic and emotional shields against the general problems of the world.

These are issues which we feel have by no means been resolved. All we are trying to do here is to set the stage for the discussion which follows. To quote Berger once again: "To some extent, definitions are a matter of taste." But if the cry is for a definition of religion as used in this study, or at least the conceptual framework within which we are working, we will have to settle for the following, which appears to be the least offensive from our perspective, although borrowed as an amalgamation of Spiro, Geertz and Berger:

> a system of symbols institutionally organised consisting of super-empirical references which establish powerful, pervasive and long-lasting moods in men by formulating conceptions of a general order of existence, and that point beyond the domain of our 'natural' reality.

The character of sects and how they relate to the study

A sect is a particular example of a group in society which, by virtue of a reactive ideology or distinctive set of beliefs or ideology, marks out a distinctive career for itself both in terms of organisation and structure and also in terms of the behaviour of its members in relation to its special ideology. One of the secondary considerations of this study is to try to suggest that such a term is not to be limited to purely religious groups, but refers to *any* group which emerges in society which takes a distinctive stance in relation to a particular belief system. It would be wrong, however, to use the term liberally in relation to any sociologically designated fad or social movement. Instead, the term is applied specifically to value-orientated groups which possess a distinctive ideology or belief system. Moreover, as most work in this area has been concerned with religious sects, it is more than likely that this will cast particular light on the nature of sectarianism in general.

The contributions of political philosophers to the study of sects has also been quite significant. Both Marx and Engels were among the first to concern themselves with the nature of both religious and political sects in the hope that a study of the former might cast light on the formation and perpetuity of the latter. Several political sociologists, mainly in the collective behaviour tradition, have stressed the sectarian phase as a necessary prerequisite for *any faith, either secular or transcendental*. Part of the close connection between the study of religious and political sects is the belief that religious sects are, according to the Marxist tradition, "prepolitical" organisations.

A full development of the term "sect" within the political Marxist tradition can probably be attributed to Trotsky and his follower Cannon, who certainly appear to have prepared the ground for subsequent social science research into the nature of sectarianism. Whatever the particular pattern of development, however, it is relatively clear that the term cannot be confined to a purely religious phenomenon but can be equally applicable to political and other phenomena which exhibit these characteristics.

There exist quite a number of sect-typologies. For example Yinger [31] talks of the *established* sect which has existed for some time but still retains an opposition to certain major issues. As for *sect* itself, he discusses three sub-categories, those that *accept* the social order, those that are *aggressive* to that order, and those that employ a policy of *avoidance*. Perhaps a more systematic attempt is that of Bryan Wilson [32] who distinguishes seven types of sect in terms of the way they respond to the world. Thus he talks of *conversionist* (evangelical and fundamentalist), *revolutionary* (anticipation of the passing of the present social order), *introversionist* (retirement from the world), *manipulationist* (claim to special knowledge and techniques), *thaumaturgical* (personal spirituality), *reformist* (provision of an ethic or critique) and *utopian (withdrawal)*. It is useful to bear in mind even at this juncture that typologies, although obviously useful to the sociologist of religion, do not have an external reality in the sense that they "exist out there". Some have suggested, also, that they tend to reduce quite complex movements and patterns of behaviour to a collection of psycho-social attitudes of an apparently simple nature. Thus Wilson's typology is based on a single and allegedly predominant number of characteristics connected with what sects appear to *do* or have as their *mission*, that is, a rather simplified picture in terms of motivations and goals. While this might be of some use in collecting together the diverse number of sects into a relative coherence, it does not really *tell* us very much. If we turn to the characteristics of religious sects, we will find that these have been relatively well documented by Wilson [33] who lists them as (i) voluntary association or joining; (ii) proof of membership, such as some special knowledge of doctrine or conversion experience; (iii) emphasis on exclusiveness and expulsion of those who deviate; (iv) self-conception of an elect who possess some special enlightenment; (v) an emphasis on *some kind* of personal perfection; (vi) a priesthood of all believers; (vii) democratic rather than hierarchical participation; (viii) room for the expression of spontaneous commitment; (ix) indifference or even hostility to the external world; (x) there is a totality of commitment; (xi) possession of a total ideology which exhibits a degree of *control* over the members' response to the external world. The cohesiveness and continuity of sects stems from the broadly common historical background of sect members and from the subsequent common moulding of sect member's lives by the adoption of a shared system of ideology. Because they know that what they believe is *right* then this only further reinforces a common pattern of action. Because they live in the real world sect members carry into that world the same problems and the same fracturing of "normality" which prompted their sect affiliation in the first instance. Thus members of the SPGG (the Socialist Party of Great Britain) and A.A. (Alcoholics Anonymous), the

Irvingites and the Swedenborgians, are respectively united outside the boundaries of their groups by the very factors which were conducive to their joining in the first instance.

This is not the place to examine in detail the reason why individuals become affiliated to sects. We refer later to various theories of participation or affiliation which we designate under the two broad categories of strain and interest. Traditionally, theorists have tended to argue that new religious movements originate as sects which can be identified as break-away movements from the established church, generally manifested as units of economic deprivation, and which eventually themselves become churches.

An attempt to account for affiliation to both secular and religious sects by other than simply economic deprivation is made by Glock and Stark.[34] They extend the definition of deprivation to all the ways in which individuals or groups may feel, or actually be, disadvantaged "in comparison either to other individuals or groups or to an internalised set of standards". Thus "religious resolutions are more likely to occur where the nature of the deprivation is inaccurately perceived or those experiencing the deprivation are not in a position to work directly at eliminating the causes. The resolution is likely to be secular under the opposite conditions where the nature of the deprivation is correctly assessed by those experiencing it and they have, or feel they have, the power, or feel they can gain the power, to deal with it directly". Several secular and religious movements have arisen which enhance both the individual and the group status of individuals, perhaps a good example being the Townsend Movement and the American Temperance Movement. Gusfield says of the latter [35] that "temperance was one part of improving status was evident in several changes in the movement during the 1830s and 1940s. In this period it changed from one in which members aimed at their own reformation". The Washingtonians, who we mention briefly in connection with Alcoholic's Anonymous, were typical of a branch of the Temperence Movement which was increasingly paying attention to the importance of sobriety and temperence as an indicator of social status.

The tendency, therefore, is to see deprivation as consisting of many kinds, other than economic, and Glock and Stark list economic, social, organismic, ethical and psychic deprivation and relate these "felt" deprivations to appropriate types of religious groups. The problem with this sort of scheme is that it is too ambitious because it tries to establish a one-one correlation between the deprivation and the group.

Both secular and religious ideologies purport to some extent to control the situation as it is. Both are essentially constraining their followers to a total cosmological stance which forces them to adopt appropriate patterns of social behaviour.

Although it is now quite common to view sects as agencies for escaping from the exigencies of the world, a refuge for those alienated from the social order which exists, this cannot be the only answer because not all who are alienated find a refuge in religion or even sectarian activities. In an attempt to overcome such a difficulty, some have argued that sects are basically a refuge for those who experience educational deprivation. It might be even more fruitful to describe sects as the homes of those who suffer some personal or social impairment in relation to the world. But again there is a deficit in this interpretation because sects almost automatically become regarded as third-rate resources of experience and pseudo and surrogate religions as in some way deviating from the "one true" church.

When we discuss deprivation, we are in the area of inquiry which views a value-orientated belief system as a mediator or bridge between social structure and consciousness. Thus "the latent resentment against society tends to be expressed in an ideology which rejects and radically devalues the society...for those in the movement, the society is symbolically transformed while actually, of course, it is left relatively untouched".[36]

Conclusion

If we return to Wilson's characteristics of religious sects, we see that all six of our groups exhibit more or less all of these. All six exhibit a pattern of voluntary joining, although surprisingly this is perhaps less noticeable with the Swedenborgians and the Irvingites than it is with the secular groups. Proof of membership in the sense of some special knowledge results in the claim by the SPGB that they are the custodians of the "true" socialism, and the prerequisite emphasis by the members of A.A. and the Swedenborgians that joiners must, respectively, relinquish alcohol and give special reverence to the writings of Swedenborg. All six groups do, and have, exercised sanctions against those who deviate from the main tenets of the group. All six groups too, entertain a self-conception as an elect who possess some special enlightenment, and all place an emphasis on some kind of personal perfection - whether living socialism as a "true" socialist, attaining sobriety through the group's teachings and methods, or a special perspective on Christianity through the media of the sect's doctrines. Certainly the SPGB exercises a claim to be a "priesthood of all believers" although the Irvingites and Swedenborgians have a hierarchy of offices which differs markedly from the democratised structure of A.A. Nevertheless, the two religious groups readily stress a "priesthood of believers in Christ". Further, commitment is essentially immediate and spontaneous in all six, although participation is more democratic in the secular groups than in the more hierarchically organised religious groups. All six groups express an indifference and in the case of the SPGB, hostility to the external world. This "spilling over" of the ideology of these groups extends to their adoption of appropriate patterns of social behaviour when living in the "real" world.

NOTES ON CHAPTER ONE

[1] For example, Roland Robertson, The Sociological Interpretation of Religion, pp.34-42. Here the problem arises out of the larger issue of definition. Colin Campbell's, Towards a Sociology of Irreligion, London: MacMillan Press, 1971; pays some attention to problems arising from definition. See also Susan Budd, "The Humanist Societies: The Consequences of a Diffuse Belief System", in B.R. Wilson (Editor) Patterns of Sectarianism, London: Heinemann, 1968, pp.373-406.

[2] John MacMurray, The Structure of Religious Experience, London: Faber, 1936, p.14f.

[3] Peter Berger and Thomas Luckmann, "Sociology of Religion and Sociology of Knowledge", Sociology and Social Research, XLVII, No.4, 1974.

[4] Phillip E. Slater, Microcosm, New York: Wiley, 1966.

[5] Max Weber, The Sociology of Religion, trans. E. Fischoff, London: Methuen, 1965, p.1.

[6] David Martin, The Religious and the Secular, London: Routledge and Kegan Paul, en passent, 1969.

[7] Ibid. p.56.

[8] Allan W. Eister (Editor), Changing Perspectives in the Scientific Study of Religion, New York: Wiley, 1974, pp.1-11.

[9] Ibid, p.5.

[10] Clifford Geertz, "Religion as a Cultural System", in M. Banton, (Editor) Anthropological Approaches to the Study of Religion, A.S.A., Monographs, 3, London: Tavistock, 1966, p.4.

[11] Melford E. Spiro, "Religion: Problems of Definition and Explanation", in M. Banton, op.cit., p.96.

[12] Robert Bellah, "Religious Evolution", American Sociological Review, 29, p.358.

[13] Will Herberg, "Religion in a Secularised Society: The New Shape of Religion in America", Review of Religious Research, vol.3, No.4, 1962.

[14] John MacMurray, op.cit.

[15] Thomas Luckmann, "Belief, Unbelief and Religion", in The Culture of Unbelief, edited by Rocco Caporale and Antonio Grumelli, California: University of California Press, 1971, pp.21-38.

[16] Peter Berger, A Rumour of Angels, Allen Lane, Penguin Press, 1970, p.70.

[17] Berger and Luckmann, op.cit.

[18] There is a vast range of anthropological literature on this area.

[19] See E.E. Evans-Pritchard, Nuer Religion, Oxford: Clarendon Press, 1956; J. Goody, "Religion and Ritual: The Definitional Problem", British Journal of Sociology, Vol.12, 1961, pp.142-164; Robin Horton, "A Definition of Religion and its Uses", Journal of the Royal Anthropological Institute of Great Britain and Ireland, 90, 1960, pp.201-226.

[20] David G. Mandelbaum, "Transcendental and Pragmatic Aspects of Religion", American Anthropologist, Vol.68, 1966, pp.1174-1191.

[21] S.F. Nadel, Nupe Religion, London: Routledge and Kegan Paul, 1954, pp.1-8, 259-278.

[22] Ibid.

[23] Peter Worsley, The Trumpet Shall Sound, London: MicGibbon and Kee, 1968, pp.21-35.

[24] Peter Berger, "Some Second Thoughts on Substantive versus Functional Definitions of Religion", Journal for the Scientific Study of Religion, Vol.13, No.2, 1974, pp.125-133.

[25] Ibid, p.128.

[26] G. Simmel, Die Religion (Die Gesellschaft, Sammlung socialpsychologischer Monographien, edited Martin Burer, Vol.X), Berlin: Marquant, 1906; translated as Sociology of Religion, by C. Rosenthal, New York; Philosophical Library, 1959.

[27] Erving Goffman, The Presentation of the Self in Everyday Life, New York: Anchor Books, 1959; Encounters, Indianapolis: Bobbs-Merril Co. Inc., 1961; Interaction Ritual: Essays on Face-to-face Behaviour, Chicago: Aldine Publishing Co., 1967.

[28] Thomas Luckmann, The Invisible Religion, New York: MacMillan, 1967.

[29] For a useful discussion along these lines see Harold Fallding, Sociology of Religion, New York: McGraw Hill, 1974, ch.1.

[30] Hans Mol, "Charisma and Conversion in the Light of Identity Theory", unpublished paper presented to the VIIIth World Congress of Sociology, Toronto, 1974.

[31] J. Milton Yinger, The Scientific Study of Religion, London: Collier MacMillan, 1970, pp.266-273.

[32] Bryan R. Wilson, "An Analysis of Sect Development", American Sociological Review, 24, 1959; Bryan R. Wilson, Religious Sects, London: Weidenfield and Nicolson, 1970.

[33] Ibid. Also his article "Typologie des sectes dans une perspective dynamique et comparative" in Archives de Sociologie des Religions, 8, 16, 1963, pp.49-63.

[34] Charles Y. Glock and Rodney Stark, Religion and Society in Tension, New York: Rand McNally, 1965, pp.242-259.

[35] Joseph R. Gusfield, Symbolic crusade, Illinois: Illinois Press, 1963, p.46.

[36] Glock and Stark, op.cit., p.248.

2 The structure of management groups: a model of the organisation and structure of resocialisation groups and settings

Introduction

Possibly the basic problem we are looking at could be described in terms of the balance that men strike between the constraints that operate upon society and the element of individual negotiation and meaning with which they interpret the world in which they live. The movements under discussion are articulated responses to the struggle for both public and private meaning in a universe which to the adherents of these movements would otherwise probably be without such meaning. But the activities in such movements are limited to the available possibilities of human behaviour and by the means and goals of the collectivities concerned. Such behaviour, that is the way in which such settings function as formally organised agencies for the socialisation, resocialisation or management of individual identity, parallels itself in any number of overtly dissimilar collectivities.

We mean by the "management of identity" the manner in which identity is controlled in a way that is in keeping with some overall formulation or set of precepts. Basically, the social interactionist perspective views an individual's identity as a result of social experiences "intrinsically associated with all the joinings and departures of social life", and such identities are social because they emerge and are sustained by a process of negotiation through interacting with others. Thus what one means by identity, according to this perspective, is that one is situated within a series of social relations and placed as a social object. In opposition to this view is the one which traditionally has regarded identity as something unaltered and fixed, a kind of given potential.

The sustaining and modification of identity, established by social relationships and processes, is to a large extent a determinant of the social structure, but as Berger has said, this is a dialectical process in the sense that such identity does itself act upon the social structure "maintaining it, modifying it, or even reshaping it". Goffman talks of "identity kits", which can be viewed as accessories to the presentation of the self, and identity can be changed or managed by covert and overt manipulation of such kits whereby the personal front is altered in some way. Thus the initiatory procedures in Synanon and Daytop, the "stripping down" of the old identity in Alcoholics Anonymous and certain T-groups, are all means which attempt the creation of new identities for the joiners. This is particularly stark in total institutions such as prisons when social and individual identity is drastically broken down in an attempt at remodelling.

Identity is not simply, either, something which hides behind one of Goffman's many masks which are worn for the purposes of impression-management, for we are not in a position to talk in terms of core-performances of identity. We are what we socially are, which in turn we reveal through our social performances. Berger talks of the manner in which our identity can become altered through the process of transformation or alternation, a process which requires appropriate social and conceptual conditions, in particular the availability of relevant "plausability structures". In itself alternation or conversion experience is "nothing much. The real thing is to keep on taking it seriously..." Thus our groups manage the identity of adherents by providing plausability structures which sustain and reinforce the initial alternation of identity. Legitimating apparatus is consequently brought to bear on the adherent, an example of which is pre-alternation biography which is

> typically nihilated in toto by subsuming it under a negative category occupying a strategic position ..."When I was living a life of sin", "When I was still caught in bourgeois consciousness", "When I was still motivated by these unconscious neurotic needs". The biographical rupture is thus identified with a cognitive separation of darkness and light.

In this way the management of identity takes place in settings which are not only plausability structures, but which also supply a cosmological or total world explanation, and which can be either secular or transcendental in nature.

Consequently, the groups used as examples of formally organised socialisation settings will be limited to value-orientated groups in the sense used by Clyde Kluckhohn, and will refer to general, organised and existential value notions which influence behaviour by being concerned with substantive claims about the universe, society or the individual. Value-orientations can be implicit or explicit, and held by groups and individuals. They are transcendental if such substantive claims refer to a super-empirical realm or secular if they refer to an empirical one. Often such groups make claims about the universe, society and the individual, for example a religious sect such as Christian Science. Some, like Alcoholics Anonymous, have as their primary concern the eudokias or well-being of the individual, while some, such as the Socialist Party of Great Britain, have as their major concern the society at large.

We are suggesting that groups concerned with managing the identity of their members in some way, whether concerned with secular or transcendental value-orientations, will exhibit similarities of organisation and structure. Such similarities in the organisation of activity is not surprising, for although individuals negotiate their own personal meaning-systems, they do so for the most part as members of collectivities or groups, with the consequence that such behaviour is not arbitrary but the product of societal processes and institutions through which individuals have passed or are passing. Groups, to be recognisable as such, require their members to limit their activities within certain areas and to particular recognisable patterns. For a group to be concerned with a transcendental or secular ideology or value-orientation (and there is a substantial overlap with these two concepts) merely stipulates a further limitation on the activities of individuals.

The structure of management groups: the work of Wheeler

Groups concerned with the management of identity (people-processing or socialising organisations) are structured, formally organised socialisation settings whose principal purpose is to change their membership in some way. Sometimes we refer to these processes as those of secondary socialisation because they extend a considerable distance beyond the socialisation processes that occur in family settings. It is by such means that individuals internalise facets of the social order. Such groups that are concerned with getting their members to internalise specific value-orientations include religious sects such as the Society of the New Church, the Catholic Apostolic Church, political sects such as the Socialist Party of Great Britain or certain revolutionary messianic groups, and secular groups such as Alcoholics Anonymous, Recovery, Inc., and Neurotics Nomine. Such organisations contain socialisation agents or persons who consciously or unconsciously have as their task the changing of the individuals (or the recruits) who enter these groups.

Groups which process (or socialise) people are integral parts of a larger processing system. Generally, they deal with input, throughput and output procedures, by which time the product, whether animate or inanimate, is processed sufficiently to satisfy the formal criteria of the group itself. Wheeler, whose work on large-scale bureaucratic organisation lends itself to the present analysis, lists a number of points or dimensions for constructing a typology of processing systems, for example the degree of differentiation of the processing agency; the nature of the object being processed (whether social or non-social).[1]

The differences between non-social and social objects of processing are obvious, and include the possibility with the latter of goals and influences outside the procedure which will militate against the group programme. The more highly differentiated the more likely the chance of conflicting roles, patterns of communication, etc., will occur,[2] and the more likely the pattern of relationships and the complexity of role interaction will be.[3] If we regard such groups as sects as relatively complex we may be aware that "recruits may be influenced in fundamentally new ways by utilising the organisation itself as a mechanism of change".[4] Wheeler gives the example of therapy groups which originated as emergency measures to cope with large numbers but which subsequently became methods of producing alternative behaviour in clients. This results in the diversity of problems which themselves have to be overcome by the structure of the organisation.

Using the method of analysis instigated by Wheeler we can view the recruits to the groups we are dealing with as attempting to cope with the management procedures which the particular groups utilise. The recruits themselves evolve workable models of the interpersonal settings they encounter. However, the groups also have their own models of the settings. Generally speaking the groups we have been describing involve entry on an individual basis into a serial process, i.e., there is a precedent pattern to follow. Obviously with the groups we are looking at the individual/serial process contains the key to the whole socialisation agent/recruit system. Initial commitment in the individual/serial process has to be sustained in ways which are alternatives to those one would use in other interactional settings.

Initial commitment and realisation of the official aims of the organisation give ample opportunity for many intervening variables, both from within and without the organisation. Recruits to socialising groups in general differ according to whether recruitment is voluntary, whether the settings are total or partial, and so on. The groups we are discussing are what Blau and Scott [5] described as "service" organisations because they are primarily concerned with the recruit himself. Because such organisations deal with correcting deficiencies Wheeler designates them as resocialisation systems. The Catholic Apostolic Church, for example, is concerned with remedying certain theological deficiencies inherent in the Christian faith, apart from its own particular slant. Alcoholics Anonymous, Recovery Inc., are both specifically concerned with the correction of certain aspects contained in the initial socialisation process. The alternative approach is through the developmental socialisation system whose prime task is to to educate and train its recruits. There is, as we would expect, considerable overlap between these two main types.

How do organisational goals relate to the social structure of socialisation agencies? In organisations such as the ones we are describing specific goals are formulated. With A.A., for example, the main purpose of the organisation appears simple and straightforward enough; to help the alcoholic obtain sobriety, and furthermore the members are generally agreed on these aims. The organisations can, of course, also differ in interpreting their socialisation as specific or general. The distinction is basically that between training (specific) and education (general). Within the organisations we are looking at it can be seen immediately that there is considerable difference in the degree to which individual members can change the system. Recruits also vary considerably, for example in sex and age, social background and so on. Alcoholics Anonymous recruits would tend to be more middle than working or upper class. Interaction, too, may be vertical or horizontal, sparse or plentiful.

In the process of movement through the structure some organisations have clearly defined patterns of entry procedures. The "conversion" procedures of value-orientated movements, for example, exhibit roughly the same stages - strain, anxiety, marking of agents responsible for anxiety state, sense of social disharmony, belief in regeneration of values, regeneration, restoration, destruction of responsible agents, belief in omnipotence of regeneration.[6] The degree of severity with which they view the entry depends to some extent on the commitment level of the recruit. Formal indoctrinatory procedures presume a high level of this commitment; less obvious commitment will require special programmes. Also, when the recruiting procedure and the initial experiences are alien to the recruit then in this case, also, formal social programmes may be utilised.

Both developmental and resocialisation agencies or organisations can vary in the amount of commitment to organisational goals exhibited by their members. In certain types of resocialisation organisations it is altogether absent, e.g., a prison. On the other hand, it is total in value-orientated resocialisation movements, or at least some of them have the character of totality. Wheeler is very unclear about this point [7] and fails to distinguish degrees and types of commitment in these organisations. However, he does make the point that recruits into developmental organisations on the whole share a knowledge of prior experience whereas recruits into resocialisation settings are essentially 'loners'.[8]

Supervisory procedures in developmental organisations differ enormously from those in resocialisation organisations. In the first type the recruit participates in the organisational decision procedures, while in the second type it is the organisation itself which makes the decisions, the recruit having abrogated this right by his failure in terms of conduct.

Another point Wheeler makes [9] is that "Developmental organisations get rid of their failures. They flunk, are expelled, or drop out. Resocialisation settings frequently get rid of their successes". Some organisations such as A.A. or certain types of religious sects use the failure-type to resuscitate members likely to stray by reminding them that they were once like this. Organisations such as A.A. and Recovery Inc., have evolved an elaborate apparatus for dealing with "backsliders" or "scapegoats".

Length of stay in resocialising organisations differs according to the goals that specific organisations have. Religious sects constantly reinforce their members' beliefs and in fact people, in theory, never leave some of the organisations concerned. Ex-patient organisations extend the time in which the recruit is in contact. Length of stay is an interesting determinant for quality of sociability, and it might be postulated that on the whole the longer the "stay" the greater the degree of sociability, and the greater the degree of stability of relationships, etc.

Wheeler discusses the variability of exit procedures and demonstrates that "different organisations follow widely varying patterns".[10] In certain instances, such as the release from mental institutions, certain safety organisations emerge to help buttress the individual against immediate contact with the outside world. Halfway houses are examples of such buttresses. Near the end of a successful completion some socialising institutions reward the recruit appropriately. Other organisations, principally resocialisation systems, place a greater degree of stress on entry rather than exit procedures.[11]

Socialisation into the roles appropriate for the organisations with which we are concerned is determined to some extent not only by such factors as the goals and personnel of the organisations but the broader social system in which the organisations exist. The organisations under discussion are constantly changing their responses to the world and subsequently the ideology of the organisations changes also. External groups can considerably modify the settings of the organisations concerned with specific forms of resocialisation. This modification can be done in a number of ways, for example by positing alternative and more attractive ideology systems. For example, A.A. procedures may be more attractive to potential recruits than a Temperance Society programme.

So far we have suggested that processing systems differ according to the degree of differentiation of the processing agency and whether the product which is being socialised is social or non-social. If it is the former then this introduces a wide range of negotiating, interacting response mechanisms. Recruits will develop meaningful definitions of settings both from their own background and from the way in which they view the settings themselves. Very often the kind of groups with which we are concerned are to be distinguished by their <u>totality</u> of commitment which is either <u>anticipatory</u> to actual joining, or is carefully cultivated by such groups immediately after the initial entry. Some difference exists, also, in the extent to which the behaviour of recruits is changed. Extreme examples are certain T-groups and psychoanalysis, but even communities for ex-drug addicts and alcoholics, such as Synanon and Daytop, and Alcoholics Anonymous and Recovery Inc., are as much concerned with the destruction of the old Adam and the birth of the new, as are groups with a transcendental ideology.

Recruits will differ because of their socioeconomic, sex and age positions in the lifecycle, and also, according to Wheeler, a) the extent to which emphasis is placed on, or a clear separation made, between Bidwell's distinction of role socialisation (preparation for specific tasks) and status socialisation (preparation to occupy a generalised "status in life"); b) the extent to which the needs of recruits mesh with those of the external community; c) the degree of <u>fit</u> between the requirements of the group and those of the individual. The degree to which recruitment is voluntary will affect the process of socialisation and so, also, of course, will the extent to which recruits are permitted to interact with others who are not recruits or committed in any way to the group's value-orientation. Wheeler also draws attention to differences which might arise from groups with a <u>homogenised</u> setting (those that tend to reduce the importance of prior experience) and <u>differentiated</u> settings (those that emphasise prior experience). The extent to which cliques or subcultures are encouraged or discouraged will also produce differences in the structure of formally organised socialisation settings.

Finally, the groups we are concerned with are composed of individuals who themselves are subject to change (if only maturational) and, again, the groups may have ideologies which change (for example, Owen Pence's study of the YMCA and its adaptation to social need, and also Zald and Denton's study of the YMCA's shift from evangelism to general service). Such groups are dynamic and interactive rather than constant and static.

We want to turn now to a duscussion of problems associated with trying to analyse the organisation of a group's structure, including meaning, pattern and boundary.

Organisational socialisation procedures are the processing agencies through which the individual adopts the organisation or group's views. It commences "whenever an individual associates with others 'under' a common symbolic label referring to that association".[12] The formal analysis of such behaviour is concerned principally with the social structure and the manner in which the actor's social role "slots in".[13]

The difficulties inherent in this formal approach are many. If we regard formal organisations as manifested for the realisation of specific goals, and possessing clearly demarcated status structures, rules, and lines of communication, it soon becomes apparent that goals are not always as obviously explicit as some suggest. To attribute goals to organisations of any kind is doubtful exercise except where there is general agreement concerning these goals by members of the organisation concerned. Talking about the goals of such organisations as the school, for example, can lead to enormous difficulties. On the other hand, original goals can be stated in a charter, but in turn become displaced by current goals, and goals of individuals may not only differ from each other but also from the original goals stated in a charter and from the current goals assumed by the organisation's leaders.

Organisational mappings of structure are to be distinguished from "normatively organised" behaviour [14] and one can view organisational socialisation from the viewpoint of "lower" as well as "higher" participants. The rules of the organisations are to be understood contextually, that is from the everyday-commonsense view of the actors.[15] "A study of an organisation must then begin with the study of its use by actors. Becoming a socially sanctioned member of an organisation, by implication, involves displaying, adhering to, and being sanctioned for use of an ordering scheme or subschemes."[16]

The static "as if" model of analysis of organisations assumes that the actor is acting "solely to the normative demands made of persons occupying their positions in a social system", but it is wrong to assume, as Goffman shows,[17] that normatively structured systems are the reality, and that knowledge concerning occupational categories tells us about the interaction process in which anyone occupying that category engages.

Manning suggests a more dynamic and fruitful approach for the analysis of organisational roles. It is the setting or group that provides "the context within which roles are assembled, maintained or dissembled". It is the settings which provide the roles with their meanings, just as it is the sentence which provides the word with its meaning. As Wittgenstein said, it is given that language is living, and it is the way that words are used, i.e., the setting, which give them their meaning.

> Roles are claimed labels, from behind which people present themselves to others and partially in terms of which they conceive, gauge, and judge their past, current and projected action. And roles are imputed labels toward which and partially in terms of which people likewise conceive, guage and judge others' past, present and projected action.[18]

We erect continuous typifications of the roles of others which we use as yardsticks to measure and construct our own typifications.

The reciprocal quality of organisational settings leads us to the view that social organisation is partly at least a process of discourse, and partly a process of signification, marking, indexing and symbolisation.[19] Organisational discourse is made sense of and understood by the actors, and only gradually by the newcomers. The peculiar language of a transcendental resocialisation setting or a mental therapy group or A.A. is not all that has to be learned by the entrant. He also has to pay particular attention to the rules which make the use of the language possible.

Meaning

Organisational categories of human behaviour are inferred by some social scientists such as Skinner from the observation of frequencies of overt behaviour. Others, such as Schutz, are more concerned with the intentions of individuals; it is by the analysis of motives that we can assume why people act as they do.[20] Weber asserted two kinds of verstehen. The first he called direct observational understanding "of the subjective means of a given act as such, including verbal utterances".[21] Here Weber is including all overt behaviour which is publicly observable, including behaviour that we infer from facial expressions. The second kind of understanding he terms explanatory understanding. "Thus we understand in terms of motive the meaning an actor attaches...in placing the act in an intelligible and more inclusive context of meaning." However, mental concepts and motives are not "inner states" or "private worlds" which affect the way we behave but are part of the description of that behaviour. There is no necessity to hold that the only means by which we can understand or know the motives, purposes of others, etc., is by a pojection of our own states. On this view it is a mistake to distinguish as does Schutz between overt behaviour and intentions, and to dichotomise, as does *verstehen*, between objective and subjective states. Motives and purposes are teleological explanations of behaviour and not inner entities which precede that behaviour.

Behaviour in social organisations is partly carried through by the meanings attached to behaviour, both by the individual to his own behaviour in relation to others, and by others in relation to his behaviour. This is multiplied further as each member attaches meaning to the behaviour he perceives. Moreover, meaning arises from the configuration of the actor, and is best understood as a function or relation to human interaction. The meanings people give to situations determines to a large extent their behaviour in those situations; and as these meanings themselves change, so their behaviour changes also.

Partly, meanings reside in the groups themselves, both in their larger settings, i.e., society itself, and in the elaborate or semi-elaborate system of offices and statuses, and ideology which reside in them. Thus, social organisations are not static entities but "territories of interaction that are focused around complex and unspoken languages".[22] These languages themselves contain the meanings which become the repositories for consequent action prescriptions.

> The organisation is...conceived in terms of competing perspectives and social orders, each of which rests on its own language and set of meanings...An organisation now becomes a multitude of shifting and competing languages and social orders, a social organisation that is held together, if at all, by a few very salient symbols such as university X, or mental hospital Y.[23]

Pattern, structure and boundary in organisational groups

Clyde Kluckhohn[24] suggested that the search for universal categories of culture has occupied various branches of the social sciences for some considerable time. Basically, the search has endeavoured to try to answer two questions, according to Kluckhohn: a) Is cultural variation limited or constrained "by panhuman regularities in biology, psychology and the processes of social interaction"; b) are substantive uniformities produced by these limits and are there accompanying trends "towards similarities in form and content"? The various criteria for the universal categorisation of culture have differed considerably, both in content and value. Nevertheless, however naive the initial attempts were, they were all endeavouring to establish regular patterns of behaviour in humans "which transcend cultural differences". And so it was that-

> Durkeim, Mauss, and other French sociologists propounded their famous principles of collective representations, reciprocity, and the like. Simmel looked for social regularities of a somewhat different type. Birkhoff in his Aesthetic Measure tried to develop panhuman canons of artistic response...The psychoanalysts tried to show the universality of the Oedipus complex, sibling rivalry, and certain sorts of fantasy and symbolic process.[25]

The methodology of comparison enables the observer to organise "concretely different phenomena in consistent terms". In addition to this "human biology sets limit, supplies potentialities and drives, provides clues which cultures neglect or elaborate."[26] Certain biological "givens", such as infant dependence, are indisputable, but there is also considerable evidence that environmental pressures can produce independent parallel development. Pressures that are similar often produce reactions that are similar. Kluckholn suggests that many aspects of human life are universal, aspects such as human anatomy, infant helplessness and dependency, and so on, and that anthropology in particular has too long concentrated on the differences rather than the similarities. Similarly, in interaction processes, there are "common elements in the structuring of social action". Groups, also, appear to possess certain basic properties. The work of Jacobson and Chomsky in linguistics indicated similarities in language that are universal, and this is probably the nearest yet to establishing an analogy with the miniscular entities of modern physics.

Universal similarities (though not necessarily identities) of culture arise because the elemental problems which confront man are basically similar. "Since most of the patterns of all cultures crystallise around the same foci, there are significant respects in which each culture is not wholly isolated, self-contained, disparate but rather related to and comparable with all other cultures."[27] While not accepting all the points that Kluckhohn makes a convenient transposition can be achieved by substituting, for our purpose here, the term "social organisation" instead of culture. It would therefore follow, as we have argued, that a) variation in social organisation is constrained by the processes Kluckhohn mentions, and in particular those of the social interaction; and b) and these constraints or limits produce substantive uniformities and similarities in form and content.

"Common elements in the structuring of social action" arise from environmental pressures. Groups or organisations have boundaries, and outside the boundaries "lies the group's environment".[28] Homans goes on to say that the group's behaviour arises out of the external system and in reaction to it. The social engineering of behaviour in the groups that we are concerned with arises in response to appeals for constraint from the external system. Certainly this is true of Alcoholics Anonymous, therapy groups, and so on, and it is also true of the transcendental movements whether we ascribe to "strain" or "interest" theories.

In terms of "goals" and "means" it is true that "social control agencies continuously monitor their deviant populations in terms of the ability of those populations to meet perceived organisational needs".[29] Labelling processes in relation to religious practitioners in the U.S.S.R., where the latter are labelled as deviant, are political and ideological castigations rather than "socially validated instances of deviance". The mentally ill or ex-mentally ill, the alcoholics and ex-alcoholics, are clearly deviant labels which are socially validated. The same is not so true of purposive ideological stances as demonstrated by membership of religious or political groups, although other factors could be introduced here which would argue, for example, that members of such and such a group were members because of...That is, arguments in terms of sexual or social deprivation, and so on.

However, to talk of "means", "goals" or "needs" with regard to organisation and groups can lead us into all kinds of difficulty. A systems approach to organisations which stressed a formal approach placed considerable emphasis on the Rational or Goal Model of organistions. In time it became apparent that the formally manifest goals of organisations did not fully explain the aspects of behaviour that could be described as the informal or "irrational" characteristics which appeared to motivate man's behaviour in organisations. Not all goals are immediately recognisable or are publically acquiesced to, and even if they were it is not at once apparent that all the members of an organistion share the same <u>concept</u> of what a goal would be. Again, some goals are not specific nor even recognised by the actors.[31] There is no guarantee that goals and means are distinguishable from one another. To say that they are -

> assumes that the means are not defined in terms of the goals; where the goal is so specific that there is only one way to achieve it, there is almost no distinction between means and goals. It is easiest to distinguish between means and goals when there are many ways to achieve the goal - that is, when the goal is vague.[32]

In the sociology of everyday life people have many goals and very often more than one means for achieving them, and within social organisations goals even if clearly recognised or explicit can become diffused. Goals can become displaced or even constantly shifting, they can conflict from different perspectives within the social organisation, some can have more value than others, and some can become means to the attainment of other goals.[33] Although perhaps we speak loosely of individuals within a social organisation possessing goals it becomes more difficult to speak of the individuals, and indeed the social organisations themselves, as possessing goals,[34] except, perhaps, where these are explicit, as in the case of Alcoholics Anonymous or Recovery, Inc. Various attempts have been made to analyse the current goals of the leadership within an organisation, or to infer the goals from observing the behaviour of participants, or by paying attention to "a set of requirements or constraints" displayed by participants.[35] The problem of "needs" is frought with similar difficulties, for example are the needs felt or perceived, basic or real, shared or private, psychological or physiological, hierarchical or democratic, and so on? There have been many modifications and extension of the "means-goal" type of analysis, for example Harary's adaptation of Merton's typology, Goffman's idea of "colonisation-indifference" to institutionalised goals and ambivalence concerning institutional areas, and so on. Some of these, such as Goffman's, are valuable applications of an Action approach, but nevertheless limited because he classifies only Total and Non-Total institutions or organisations.[36]

We have to some extent discussed pattern and structure in organisation. The task which is left to us is to look at the boundaries of these groups. Boundaries are circumscriptions of events or entities in the external world, which delineate these events or entities from society as a whole. We have already suggested that entities or groups, for example the ones under consideration, are configurations of the real world upon which we have focused and identified as patterns of behaviour which we have recognised. Several criteria sprang to mind, and we mentioned a paucity in the "web of interactions"; activities recognised as being distinctive; time; membership, and so on, before coming to recognise that Denzin's phrase "territories of interaction" usefully tied in with a gradual realisation that "rules and "meanings" were heuristic devices to describe certain aspects of human behaviour.

Basic to the rules of grammar and the development of the conscious self is the notion of separateness. That is, the process of social interaction involves a continual process of selection and separation from the mass of stimuli with which we are confronted in the external world. The notion of boundary is inseparable from the notion of structure, and subsequently from the marginal areas and zones of indifference[37]

The distribution of belief structures in value-orientated movements

Value-orientated belief systems involve a reconstitution

of society with specific reference to the legitimate ends which modify social action, and are expressed in general statements. In other words, a deliberate intention is declared to refashion or reconstruct social reality. Furthermore, the objectification of beliefs are sustained by specific ideological settings which become the "carriers" and the "official legitimators" of these beliefs.[45] A value-orientated movement or organisation is a collective attempt to enforce or perpetuate these value-orientated general statements. After conversion or "alternation" to value-orientated movements such conversions or alternations need an indispensible "plausibility structure".[46] Therapeutic mechanisms safeguard the boundaries of such organisations and correct "deviant" tendencies. Re-socialisation settings of value-orientated groups provide members with a plausability structure "properly separated from the outside world and equipped with the necessary socialising and therapeutic personnel".[47]

Conclusions

Participants or recruits in the groups we are concerned with are having their identity managed in some way or another. Wheeler's typology of processing systems casts particular light on this process. Individuals enter into serial processes within these resocialisation settings, and these settings change in relation to their environment together with their ideologies. Common symbolic labels group together interactional patterns with which the actor associates. The settings or organisations provide role contexts and give the roles meanings, and consequently emerge as miniature universes of discourse. Interpretive stances towards organisations, or the meanings attached to behaviour, to a large extent determines that behaviour.

Varying value-orientated settings and groups display many of the characteristics of both secular and transcendental ideology, structure and organisation. Both categories are concerned with the management of the identity of their members in relation to a total and embracing ideology. The <u>content</u> of such ideology is in many ways less important than the behaviour within the organisation and structure of small value-orientated settings which is designed to facilitate their task - or resocialisation - in some manner. It is for this reason that identity management groups of both categories will be expected to erect similar structures because there exists no other identifiable means for carrying out or facilitating the task at hand. In other words, there exist a <u>minimum</u> of constancy of interactive patterns and symbol forms that are necessary for the groups concerned to "put over" ideological legitimations, resocialisation procedures, and so on. These total, and perhaps deviant, ideologies locate and solve to each participant's satisfaction the moral and social dilemmas of society. The management of identity in cosmological settings gives some insight into the similarities of organisation and structure between secular and transcendental value-orientated groups. Such similarities in the organisation of human activity as we argued earlier is not really surprising because as members of collectivities and groups, individuals negotiate their own very personal meaning-systems.[38] Such behaviour is far from arbitrary and is rather the result of processes in society and in the institutions of society through which individuals pass. In order to recognise behaviour as group behaviour members must not only confine their activities within certain limits but also to particular recognisable patterns. For such groups as the ones which we have discussed, concerned with secular and transcendental value-orientations and ideologies, this simply stipulates a further constraint on the behaviour of the individuals concerned.

NOTES TO CHAPTER TWO

[1] D.F. Swift, The Sociology of Education, Routledge and Kegan Paul, 1969, pp.44-47, following Orville G. Brim, and Stanton Wheeler, Socialisation After Childhood, New York; Wiley, 1966, pp.56 ff.

[2] Orville G. Brim and Stanton Wheeler, op.cit., p.57.

[3] Ibid, p.58.

[4] Ibid, p.59.

[5] Peter Blau and Richard W. Scott, Formal Organisations: a Comparative Approach, London: Routledge and Kegan Paul, 1963.

[6] N. Smelser, Theory of Collective Behaviour, London: Routledge and Kegan Paul, 1962, p.124. See also R.K. Jones, "Paradigm Shifts and Identity Theory: Alternation as a Form of Identity Management" in H. Mol, Prospects and Problems of an Identity Theory of Religion, London and Beverly Hills: Sage Studies in International Sociology, 1978; R.K. Jones, "Some Epistemological Considerations of Paradigm Shifts: Basic Steps Towards a Formulated Model of Alternation", Sociological Review, May, 1977.

[7] Brim and Wheeler, op.cit., p.86.

[8] Ibid, p.87.

[9] Ibid, p.89.

[10] Ibid, p.93.

[11] Harold Garfinkel, "Conditions of Successful Degredation Ceremonies", American Journal of Sociology, Vol.LXI, 1956, pp.420-424.

[12] Egon Bittner, "The Concept of Organization", Social Research, 32, pp.230-255, quoted in Peter K. Manning, "Talking and Becoming: a View of Organizational Socialisation, in Jack D. Douglas, Understanding Everyday Life, London: Routledge and Kegan Paul, 1971, pp.239-256.

[13] Peter K. Manning, op.cit., p.240.

[14] Ibid, p.241.

[15] Ibid, p.243.

[16] Ibid, p.244.

[17] Erving Goffman, "Role Distance", in Encounters, Indianapolis: Bobbs-Merrill, 1961, quoted in Peter K. Manning, op.cit., p.245.

[18] John Lofland, "Role Management: A Programmatic Statement". Mimeographed paper No.30 of the Center for Research in Social Organization, University of Michigan, June, 1967, quoted in Peter K. Manning, op.cit. p.247.

[19] Peter K. Manning, op.cit., p.251.

[20] Alfred Schutz, "The Social World and the Theory of Social Action", in David Braybrooke, (editor) Philosophical Problems of the Social Sciences, New York: MacMillan, 1961, p.64.

[21] Max Weber, Theory of Social and Economic Organization, New York: Oxford University Press, 1947. See also Arthur W. Diquattro, "Verstehen as an Empirical Concept", Sociology and Social Research, 1973, pp.32-42.

[22] Norman K. Denzin, "Symbolic Interactionism and Ethnomethodology", in J. Douglas, editor, Understanding Everyday Life, London: Routledge and Kegan Paul, 1971, pp.259-284, p.281.

[23] Ibid, p.282.

[24] Clyde Kluckhohn, in A.L. Kroeber, Anthropology Today, pp.507-523.

[25] Ibid, p.511.

[26] Ibid, p.513.

[27] Ibid, p.521.

[28] George C. Homans, The Human Group, London: Routledge and Kegan Paul, 1956, p.86.

[29] Denzin, op.cit., p.280.

[30] David Silverman, The Theory of Organizations, London: Heinemann, 1970, pp.26-27.

[31] Percy Cohen, Modern Social Theory, London: Heinemann, 1968, p.70.

[32] Ibid, p.71.

[33] Ibid, p.72.

[34] David Silverman, op.cit., p.9.

[35] Ibid, p.11.

[36] Ibid, p.23.

[37] S. Krupp, Pattern in Organization Analysis, New York: Holt Rinehart and Winston, 1961, pp.53-74, although in Krupp's case he is discussing the boundaries of a theory. Nevertheless, his analytical framework is, I feel, one way of approaching this problem.

[38] Burkhart Holzner, Reality Construction in Society, Cambridge, Mass: Schenkman Publ. Co. Inc., 1968. The ideologies of value-orientated groups consequently emerge as legitimating symbols and function as the unifiers of collectivities which assist in the "definition of collective boundaries, both internal and external. Often they limit the permissable range of specialised reality construction within specialised work communities by restricting the range of tolerated deviation. Positively, an ideology may aid in the mobilisation of energies for the pursuit of collective goals...(and)...serve epistemological functions by defining a collectivity's basic assumptions about the world and about society". Holzner distinguishes four types of alternative ideologies: a) loyalistically radical movements which require an ideology of resurrection or revitalisation, e.g., the Nazi Party. b) alienative radicalism adopts alternative ideologies which stress "the existing system to be unworthy of further continuity"; e.g., socialist movements. c) accommodative sectarianisms stress the limiting of area "through the erecting of sectarian boundaries and the concentration of personal virtue and personal hopes"; e.g., the Wandervogel movement among German youth. d) escapist sectarianism rejects the dominant values of the system; e.g., Old Order Amish. He speaks, interestingly, of a dialogue among the many ideologies of society, and how in a way they determine each other (they tend to be "mirror images" of each other). The foregoing is applicable to the wide range of alternatives to be found within and between secular and transcendental ideology systems and organisations.

3 Transcendental characteristics of non-transcendental groups

If, as we are suggesting, non-transcendental groups do display considerable similarity to transcendental groups we must do two things: a) reveal the nature of this similarity; b) describe its characteristics. Transcendental and non-transcendental groups may resemble each other for many reasons.

1. Secular, non-transcendental groups may appropriate the symbols of religious culture to facilitate or "act out" their own group system.

2. Religious or transcendental characteristics may be inherent in group processes as a whole.[1]

3. Transcendental and non-transcendental groups may be functionally equivalent (inclusive identification).

4. Overtly secular or non-transcendental groups are in effect religious surrogates or pseudo-religious.

5. Secularisation has, perhaps, blurred the distinction between transcendental and non-transcendental groups (if such a sharp distinction ever existed).

6. The structure of groups may well overlap, i.e., empirical systems correspond in certain respects.

7. Belief systems may be much the same and any differentiation may lie in the reference or perspective through which we view the empirical realm - we take up different "stances" to various alternative legitimating systems of the universe.

8. Apparently dissimilar movements may possess the same or similar organisation and structure.

9. There is, perhaps, a minimum of constancy of interactive patterns and symbol forms that are necessary for the functioning of certain sorts of groups.

10. There are psychological or sociological universals or "universal laws" which are both part of the "mind" and part of the mirroring of the empirical realm.

Before we proceed to look at each of the possible reasons for similarity it is imperative to make quite clear what is meant here by the assertion that transcendental and non-transcendental groups may often display similar characteristics; we need to reveal the nature of the similarity. What is being repudiated is the idea of the existence of a kind of unitary essence which is common to both. In the Meno, Socrates asserts,

> We always arrive at a plurality, but that is not the kind of answer I want. Seeing that you call these many particulars by one and the same name, and say that every one of them is a shape, even though they are the contrary of each other, tell me what this is which embraces round as well as straight, and what you mean by shape when you say that straightness is a shape as much as a roundness...What is this thing which is called "shape"?...Don't you see that I am looking for what is the same in all of them?...What is it that is common to roundness and straightness and the other things which you call shapes?[2]

The notion of a common essence was attacked most forcibly by Wittgenstein [3] when he talked about "family resemblances". He uses as an example the notion of games and denies that when we look at all the games there are we would find a common characteristic which makes us call them by the same name.

> If you look at (games) you will not see something that is common to all, but similarities, relationships, and a whole series of them at that. To repeat: don't think, but look! - Look for example at board games, with their multivarious relationships. Now pass to card-games; here you find many correspondences with the first group, but many common features drop out, and others appear. When we pass next to ball-games, much that is common is retained, but much is lost. - Are they all "amusing"? Compare chess with noughts and crosses. Or is there always winning and losing, or competition between players? Think of patience. In ball games there is winning and losing; but when a child throws his ball at the wall and catches it again, this feature has disappeared. Look at the parts played by skill and luck; and at the difference between skill in chess and

> skill in tennis. Think now of games like ring-a-ring-a-roses; here is the element of amusement, but how many other characteristic features have disappeared! And we can go through the many, many other groups of games in the same way; we can see how similarities crop up and disappear.[4]

The same point was made by William James[5] much earlier when talking about "government" and "religion", but Wittgenstein's argument is that we are not arbitrarily labelling a multivarious collection but that the multivarious collection has in common "family resemblances".

> ...We see a complicated network of similarities overlapping and criss-crossing; sometimes overall similarities, sometimes similarities of detail.
>
> I can think of no better expression to characterise these similarities than "family resemblances"; for the various resemblances between members of a family: build, features, colour of eyes, gait, temperament, etc., etc., overlap and criss-cross in the same way. - And I shall say: "games" form a family.[6]

The point that Wittgenstein is making is that even though several things are given a general term to designate them they do not possess a characteristic or set of characteristics in common - they share no common <u>essence</u>, neither do they possess a unitary characteristic. If transcendental phenomena have <u>all</u> the characteristics of non-transcendental phenomena then naturally there would be no distinction between them and we would therefore not be entitled to call one set of phenomena "transcendental" and the other not. Similarly, there must be a minimum degree of "transcendental" characteristics present in order for us to designate a set of phenomena as transcendental; and so with non-transcendental phenomena. Transcendental phenomena may have certain characteristics which we designate as a,b,c,d,n, and these characteristics, if no other phenomena had them, would then, in Wittgenstein's sense, possess a unitary characteristic. In reality, of course, transcendental and non-transcendental phenomena have only <u>some</u> of the characteristics in common. The essentialist view of phenomena would hold that such phenomena possessed a) <u>essential</u> characteristics that only group x possessed; b) <u>accidental</u> characteristics not confined or essential to the phenomena. Wittgenstein denies any such distinction.

The alleged distinction between essential and accidental properties reflects to some extent on the claim of some who maintain, like Levi-Strauss, that structure can be recognised and categorised irrespective of constituents. If we remove all the properties of an object, nothing remains, and to have any conception of an object we must take into account the constituents or properties of the object.

Reasons for resemblance

We return now to an analysis of the ten possible reasons for the resemblance of transcendental to non-transcendental groups. It may often happen, of course, that groups in category 1. may be secular in function and non-transcendental in intention. Such groups may utilise particular aspects of religious culture in order to facilitate certain goals. A nurses training school,[7] for example, may often deliberately adopt, or intentionally resemble, certain religious orders. Other groups, such as Buchmanites, Alcoholics Anonymous and so on, may adopt particular aspects of religious or transcendental groups, such as "group confession", in order to achieve more effectively certain pragmatic goals.

Category 2. has doubtful, but respectable antecedents in Jung's "collective unconscious", Freud's theory of "mind",[8] and Levi-Strauss's "psychological universals". Recent proponents, notably Slater and Swanson, have suggested a more sophisticated approach. Slater's concern is essentially with group structure, and particularly with the relationship of groups to their leaders. "First order" problems are generated by such events as death, deprivation, and isolation and these in turn generate religious ideas and sentiments. Attitudes of dependent reference emerge from situations of dependent longings, and when the real object is removed. This is particularly evident in T-groups when the leader deliberately adopts a passive role, and this passive role is experienced as a death or loss. Because the group lacks authoritarian directives and there is a minimisation of personality cues due to the passivity or silence of the leader, the leader is viewed by the members of the group as being dead or lost. The "experiment" myth is an example of a religious cosmogony in microcosm - all will be understood in the "latter day".[9] Deprivation at the hands of the leader gives way to a deification of him,[10] and research on small groups directed along these lines cites frequent "religious" references, i.e., worship, the leader as an "alter", and so on. The leader himself assumes a somewhat blurred sacred/secular quality; something akin to Frazer's "sacred king". Macroscopic religious evolution in society, Slater claims, has many parallels with the evolution of sacred attitudes in a T-group. There are several explanations which one could offer such as the psychological one stemming from Freud and Jung which suggests the existence of primordial "images" in the unconscious, though Slater finds more sympathy in a sociological explanation of a Durkheimian nature which is suggested by Swanson. This sort of explanation stresses the impact of common structural conditions on group ideation. Slater describes the kinds of changes brought about in more or less shared fantasy systems by rearrangements of the bonds linking individuals together. He suggests, indeed, that social structural changes must also be paralled by changes in the psychic structure. Slater stresses a great deal the process of boundary awareness, an idea which Piaget sees as important for co-operation, which is based on consciousness and separateness, and which one must be aware of; this awareness of boundaries enables one to communicate with others. Slater makes three important points:

a) groups regularly utilise religious symbolism, that is, there is a tendency for groups to draw upon a rich cultural heritage in expressing whatever reactions they may have to the difficulties involved in establishing and maintaining a complex set of interpersonal relationships;

b) group processes may very well not be randomised but in fact display patterns often replicated in the past: groups <u>must</u> recapitulate <u>some</u> cultural evolution;

c) the idea of boundary awareness seems to suggest "an underlying principle of human association" simply because we might well be observing phenomena which derive from patterns of human interaction which are fundamental, and such patterns describe the manner in which units of living matter combine, organise and dissolve.

Slater means, naturally, that for humans it is not organic matter which is undergoing these processes but their artifacts, symbol systems and response patterns.

Swanson's argument is a perpetuation of Durkheim's. Whereas the latter argued that religious experience arose out of man's relation to society because mankind felt that society dominated its actions, sanctioned its behaviour, and exists before and after it, Swanson identifies the supernatural with particular groups in society which "persist over time and have distinctive purposes". What he maintains is "that people experience 'supernatural' properties in social life not merely because men are unwittingly controlled by social norms which they learn, but because social relationships inherently possess the characteristics we identify as supernatural".[10] Certain relationships or groups, which he designates as "sovereign", perpetuate in their interaction an enduring set of relationships called "constitutional arrangements" or "constitutional structures". These constitutional structures give a defined statement of procedures and are what men "often conceptualise as personified and supernatural beings".[11] The characteristics we attribute to supernatural beings approximate most closely to those characteristics possessed by the primordial and constitutional structures of social relationships. These social relationships overwhelm and govern our inward and outward experiences, and give us purposive directions for action. In other words, there are certain characteristics in social interaction and social life which, according to Swanson, correspond with the supernatural elements or our conception of these elements. He makes the point that there are important distinctions between transcendental and secular movements, bodies, groups, etc:

> We blur important differences in belief if we say that magic is only the science of primitives or that social movements like Fascism or Communism are of the same cloth as Christianity or Buddhism. There are similarities between magic and science. Both, for example, seek to understand and control nature. Likewise, both secular and religious movements may embody values which men seek with great devotion. But the underlying philosophy of science ignores the supernatural, and secular creeds, as such, either ignore supernature or deny its existence. Magic and religion remain distinctive in having contacts with the supernatural as their goal.[12]

Swanson means by the term "sovereign organisation" types of organisations which "exercise original and independent jurisdiction over some sphere of social life".[13] This jurisdiction is independent if no other organisation or individual can legitimately supplant the decisions which it makes; the jurisdiction is original if no other organisation or individual can legitimately initiate a decision in the sphere of social life. He has in mind organisations or groups such as the family (nuclear) and _gemeinschaft_ settlements. Swanson proceeds to elaborate a set of criteria for formal qualification as a sovereign group. The constituent members of a group or those that represent it should meet at least once a year; there must be institutionalised or customary procedures for decision-making, and legitimated formal or informal roles; the group's goals and means must be regarded as legitimate by its members, _generally speaking_; the group must be regarded by its constituents as existing into the future; the group must have three or more members; its decisions must have a significant effect on its constituents; it must be independent, and not an agency of another organisation; it must be regarded as a distinctive organisation by its constituents. In other words, what Swanson is saying is that if any group or organisation fulfills these criteria it is a sovereign group which _inherently_ possesses the characteristics which we designate as supernatural. He appears to be suggesting that we explore a particular task, experience an aspect of behaviour and, discovering we like it, we subsequently engage in it yet further until it becomes a definite pattern or part of the "constitutional structure". These patterns are enduring, and persist over time. They provide a "continuing, if subterranean, set of forces which orders many of one's activities". These dispositional patterns forge the primordial links between men. Both primordial and constitutional structures have properties which Swanson defines as supernatural.

Both Slater and Swanson present a cogent thesis. The evidence which the former cites for a substantiation of his thesis is largely psychoanalytic in nature and consequently difficult to verify. It consists of largely personal, subjective accounts of experiences or interpretations of experience. Slater's evidence is largely drawn from T-groups,[14] which are not only to some extent suspect as legitimate procedures but are also, of course, vastly different _sorts_ of groups from those which Swanson talks about. It may be, perhaps, that "first order problems" _do_ generate religious ideas and sentiments, but these responses are not in any sense _inherent_ responses. They are simply _learned_ from the cultural milieu. The alleged religious responses to absent leaders in groups, and especially T-groups, is based almost entirely on a purely arbitrary _interpretation_.[15] It is not unnatural that the leader of a group composed, on the whole, of people suffering from some degree of _imbalance_, should have displayed towards himself some particular quality of relationship.

In order to achieve a set of goals a group must use either institutionalised means or at least adopt a minimum set of conditions or relationships necessary to function effectively. Groups appear to replicate past processes only because these processes have been proved effective.

Swanson's primarily sociological argument fails to bridge sovereign groups and the characteristics (i.e., being supernatural) "inherent" in the quality of the relationships in these groups. He doesn't refute Durkheim's claim that men are unwittingly controlled by the social norms which they learn, that is, that we assimilate dispositional stances, and indeed introduces the notion of inherent characteristics. The set of criteria for qualifying as a sovereign group are not substantiated by Swanson.

Category 3. maintains that transcendental and non-transcendental groups may be functionally equivalent, i.e., although apparently dissimilar they both perform the same _functions_. This inclusivist argument takes as the criteria for the indentification of two or more differentiated groups the fact that they appear to perform the same or similar functions, and in a _definite_ sense it might assert that Alcoholics Anonymous, for example, _is_ a religion, whereas in a less definite sense it might suggest that Alcoholics Anonymous is _functionally equivalent_ to religion.[16]

It is argued that if a particular pattern of emotional gratification is no longer maintained, then men must be fulfilling their desires by other means: can one define these means as less religious than those which are traditionally employed for the same ends?

The argument appears to posit a more or less universal and unchanging human psychology. It takes little or no account of the changing patterns of social organisation, social experience and social opportunity in modifying men's emotional needs and responses. It commonly accepts the functionalist argument, that religion exists to meet certain human needs. Assuming that these needs are unchanging it is faced with the fact that none the less religion changes. Having derived the needs from religious phenomena, it now refers to the needs in order to assert that religious practices must exist to meet them. Thus if a given form of religion disappears, then some other "religious" behaviour must be arising to take its place. Thus mass-entertainment, mass-rallies, science, totalitarian political parties and ideologies, expensive consumer goods are variously credited with fulfilling some of the functions once fulfilled by religion.[17]

The objection to category 4. is much the same. Pseudo or surrogate religion may be either <u>intentionally</u> so, or there may be characteristics attributed to a group which result in people subsequently designating it as such. Thus communism may be called a pseudo or surrogate religion, whereas Alcoholics Anonymous might resemble certain sectarian movements or groups. Again, some modern movements such as religio-psychoanalytical groups might reject major components of established religious organisations and yet retain some of these components, or at least enough to ensure the designation "pseudo religious".

The process of secularisation,[18] which is what lies behind category 5., is concerned basically with the retreat of the <u>religious</u> from the realm of the everyday world and the gradual innovation of the secular. It may well be the case, as some have maintained, that although institutionalised religious structures remain apparently little changed the essential religious commitment has dwindled significantly. Beliefs and values previously claimed as the province of the transcendental are increasingly subscribed to on <u>moral</u> grounds. Shiner [19] distinguished several varieties of the secularisation concept.

(a) The decline of religion means simply that religious characteristics previously accepted as legitimate have lost their prestige and influence, the result of which will be a religionless society.

(b) Religious adherents and societies become increasingly interested in this world.

(c) Religion ceases to have societal importance and becomes relegated to the private sphere.

(d) Religious or transcendental patterns are transposed to the secular sphere, e.g., psychotherapy replaces confession and the Marxist theory of revolution replaces the Jewish/Christian eschatological belief.

(e) Human society and processes are regarded less as sacral processes and more as rational "cause-effect" process.

(f) The movement from the sacred to the secular society is in essence a movement to a society where actions are expediated on the grounds of rationality and utilitariansim.

Nearly all these varieties of the secularisation concept imply some error in index or definition. For example, when we talk of a process of secularisation where are we to discover the alleged religious age from which the decline of religion has commenced? Some theologians[20], also, have advocated secularisation or commitment to the world as the only positive religious stance.[21]

When we turn to category 6. what is suggested here is that empirical systems do often overlap or correspond in certain respects. For example, it may be expected that overtly dissimilar groups may correspond to each other in such factors as degree and kinds of interaction, cohesion, institutionalised procedures, leadership types, channels of communications, status ascription and achievement, goals and means, etc. Again, groups may correspond to one another in the manner in which leaders emerge, in the tasks, in motives of affiliation, and so on. Social organisation patterns may overlap in different empirical systems. Etzioni,[22] for example, brings together several quite distinct organisations which rely predominantly on normative control, and these organisations adopt a normative structure for the acceptance of directives and means. The normative organisations, which Etzioni lists, are political, religious, educational, medical and voluntary. These are organisations where the primary means of control is that of normative power and where the organisational orientation or perspective is characterised by an intense level of commitment. Such organistions as these rely on the internalisation of directives which are regarded by the members as legitimate. He cites leadership, rituals, the manipulation of social and prestige symbols, and processes of resocialisation, as among the most popular means of enforcing social control. Voluntary associations are common interest groups displaying voluntary membership with high commitment and involvement, and they bear resemblance to religious groups in most aspects apart from an institutionalised transcendentalism.

Category 7. implies that it may be the case that belief systems, religious and ideology systems, might well be saying much the same thing about both the human situation and man's response to that situation. What one might have in mind here is the situation described by Pope[23] in his analysis of a strike which occurred in a single cotton mill in the small city of Gastonia in North Carolina in 1929. Apart from the fact that the local churches condoned the action of the industrial mill owners, Pope showed how the workers joined the Holiness sects. His point was that the religious response was only one alternative among many, and that they migh just as easily have become affiliated to a labour movement. Barber, Worsley and Lantanari[24] have between them shown that religious and political manifestations are not always readily distinguishable. The question is, of course, not simply one of differentiating between the belief systems themselves but trying to explain why some people in society opt for a religious or transcendental stance while others do not. As others have pointed out,[25] societies need to have some sort of legitimation, and in an increasingly secularised society other than religious systems of legitimation are emerging. These alternative legitimating systems "compete for the patronage of potential consumers of Weltanschauungen".[26]

Category 8. is suggesting a similar point to that of category 6., although limiting the similarity of transcendental systems to non-transcendental systems to similarities in organisation and structure. The validity of this view will, I hope, become apparent in subsequent chapters.

Category 9. suggests that there is a minimum or basic number of interactive patterns and symbol forms that are necessary for the functioning of groups, and it is therefore evident that overtly distinctive groups ought, on analysis, to reveal similar interactive forms. All groups, to some extent, possess a sociometric structure (irrespective of whether a sociometric study accurately depicts the actual situation). There is a minimal degree of attraction to the group, there are goals and means to goal achievement, there is a normative structure which consists of shared ways of interaction, there is a belief system, and so on. There are institutionalised procedures for deviants to comply. All groups have some form of hierarchical structure and role differentiation, and the former can be overt or covert, verbal or social.

Power, authority and leadership are characteristics of group structure. Comparisons of transcendental and non-transcendental groups are bound, to some extent at least, to reveal analogies in certain of these characteristics. Tuckman[27] suggested that group structure was not only related to task activity but that groups go through developmental stages. These stages are <u>forming</u>, which is a period of anxiety and uncertainty, <u>storming</u>, which is a period of conflict and sub-grouping, <u>norming</u>, which is a cohesive and mutually supporting period, and <u>performing</u>, which is a period of flexible and functional role-playing. Groups vary over the period of time taken to pass through these stages and in the degree of <u>completion</u> of the characteristics associated with these stages.

The last category, 10., postulates the existence of something akin to a "group mind", a primordial depository of images which are shared by all mankind and which perpetually reappear through myths and folk tales. These images reflect the "substance" of society and the immutable aspects of the empirical world. Apart from the difficulty in understanding what is being conveyed by this notion, its logical geography is rather strange. The positing of psychological or sociological universals is based on a crude rendering of organicism and can be found in varying forms in the work of Durkheim, Giddings, Cooley and others, although it has reappeared in a more sophisticated form in the work of Levi-Strauss. Freud, Jung and others, notably McDougal, introduced this or variations of this idea into the field of psychology. Cooley, for example, viewed society as consisting of both a psychic entity and an organic whole.

So far as Jung was concerned the collective unconscious was the sphere from which personal consciousness arose. The collective unconscious was not a storehouse for <u>inherited</u> primordial behaviour and experience but a shared collection of experiences which through time had shaped the human brain; what Jung called "pre-existent forms of apprehension". These archetypes are the result of accumulative experiences for the whole of mankind.[28] There are certain human situations in society, and symbolic forms perpetuate and even "contain" these, where we experience the archetypes. Jung, like Levi-Strauss, found myth as the chief receptacle of archetypal experience. Theories of this sort are difficult not only to comprehend but also to substantiate. What counts as valid empirical evidence and what criteria of validation are we using? For example, the common element in myth forms may well be due to diffusion and other means of cultural spread rather than to symbols inherent in the social structure.

What has so far been done, then, is to examine the various ways in which transcendental groups might resemble non-transcendental groups. The nature of the similarity is a complex one, and of the ten theories presented some have been discarded while others have been suggested as having some validity.

Accepted reasons for similarity:

a) Secular, non-transcendental groups, may appropriate the symbols of religious culture to facilitate or "act out" their own group system.

b) Transcendental and non-transcendental groups may be functionally equivalent (inclusive identification). This is accepted with certain reservations.

c) Overtly secular or non-transcendental groups are in effect religious surrogates or pseudo-religions. This is accepted with certain reservations.

d) The structure of groups might well overlap, i.e., empirical systems correspond in certain respects.

e) Belief systems might be much the same and any differentiation might lie in the reference or perspective through which we view the empirical realm - we take up different "stances" to various alternative legitimating systems of the universe.

f) Apparently dissimilar movements may possess the same or similar organisation and structure.

g) There is, perhaps, a minimum of constancy of interactive patterns and symbol forms that are necessary for the functioning of certain sorts of groups.

Unacceptable reasons for similarity:

a) Religious or transcendental characteristics may be inherent in group processes as a whole.

b) There are psychological or sociological universals or "universal laws" which are both part of the "mind" and part of the mirroring of the empirical realm.

c) Secularisation has, perhaps, blurred the distinction between transcendental and non-transcendental groups (if such a sharp distinction ever existed). This is unacceptable with certain reservations.

Retrospective research on parallel secular movements

Political groups:

Werner Stark [29] brings out some interesting analogies between political, especially communist, movements and religious ideology. He argues that the basis of monarchical religion was destroyed when society shifted from a communal to an associational pattern.[30] This reorientating of the social system also implied an allocation of "religious significance: the lccus of holiness, we may perhaps say, was shifted".[31] This meant, in effect, that "there was a replacement of an irrational-religious by a rational-irreligious system of life" and "the task of bursting open the doors of paradise was reserved for the new suffering servant, the working man".[32] He quotes from the Christmas Day, 1955 edition, of Vorwarts, the organ of the German Social Democratic Party:

> The salvation of the masses, not through a miracle-working saviour, but through the masses themselves - that is socialism's idea of salvation, that is our saving gospel...The masses of the people have gone over to self-salvation, from prayer...to fight, from the crucifix to the red banner of Social Democracy.[35]

"...the attribution of a messianic role - or rather of the messianic role - to the working classes and masses"[34] seems to have been quite prominent in Michelet's work. Stark is suggesting that Marxism has extremely strong affinities "to those truly revolutionary stirrings which rejected the established state of humanity and tried to usher in a new aeon, a new dispensation".[35] He extends his argument.

> ...Soviet messianism justifies a plan for a Russian Communist world organised and directed from Moscow...Marxism, in effect, becomes a political religion. Russia is the chosen nation, not only entitled, but duty bound to bring light and truth to the "toilers" of those parts of the world who have not yet beeen saved from "capital slavery".[36]

Stark, finally, turns to a report in the Manchester Guardian (29 March and 25 August, 1962) which deals with the Ghana Young Pioneers' Order No. 3 concerning the Institutionalisation of Osafego Dr. Kwame Nkrumah "...At all Ghana Young Pioneers' activities, e.g., at the beginning of the school day, the order runs, members should respond to the Institutionalisation Slogan and sing The Fishers of Men". The Institutionalisation Slogan runs as follows:

> Leader: "Nkrumah does no wrong".
> Response: "Nkrumah is our leader".
> Leader: "Nkrumah does no wrong".
> Response: "Nkrumah is our messiah".[37]

He also reports that Dr. Nkrumah had his mother portrayed as the madonna.[38]

The Guardian, similarly, in 5th June, 1961, summarises a speech by Oginga Odinga, Vice-President of the Kenya

African National Union, in which he claimed that Jomo Kenyatta was the Africans' "second God", and the Europeans who did not believe in that God were "devils". The healing and rain-making powers which were ascribed to President Sukarno are similar manifestations.[39]

Stark's parallels are based on general arguments and superficial appearances rather than any systematic examination of communist groups or ideology and their transcendental counterparts.

Similar general parallels are indicated by others. Heberle,[40] for example, mentions Sir John Maynard's characterisation of Lenin's conception of the Communist Party as that of

> a disciplined order of devoted adherents, more clearly resembling the Society of Jesus than any of the lax aggregations of political sympathizers to which we are accustomed to apply the name of political party...We can trace the germs of it to Bakunin, Nechayev and Tkachev.

Rossi, again quoted in Heberle, mentions the <u>totality</u> of influence which the Communist Party asserts over its members, leaving no separate category between the personal and political areas of life. "The party is, on this showing, less a party in the ordinary acceptation of the term than an ecclesia". We are reminded, also, how the National Socialists used the model of the medieval orders of religious knights - Templars and Teutònic Knights, and also the analogy between political parties and the Jesuits. Heberle suggests that the Communist Party in Russia is a new form of an ancient institution - the priesthood.[41] "The Russian religious conception of the presence of truth in the congregation, and there alone, passed to the Communists."[42]

Stiernotte[43] has suggested the similarities of aspects of mysticism and communism at the discussion of "cosmic philosophies". He brings testimony from various sources that the Marxist doctrine of dialectical materialism is essentially remarkably mystical in character. He quotes with approval from Wetter[44] that there are several points of contact between mysticism and communism. For example, dialectical materialism emphasises a sense of mystery about the world; it implies a sense of connection and unity; both Marx and Engels mentioned certain mystics as their precursors; however, he fails to substantiate or sustain these parallels. Similarly, McIntyre deals several blows to the suggestion of either analogies or reapproachment.[45]

An almost identical point to that of McIntyre's is made by Zeldin,[46] who not only explores Marxist-Christian parallels but denies the possibility of any <u>dialogue</u>

between the two unless each is prepared to relinquich its essential nature. Statements identifying communism and so on with religion are "pseudo-statements; for the terms are too vague to permit discriminate application". Admittedly, the language of the young Marx was religious but numerous writers have tried to construct a socialist religion.[47]

> Marxism has a doctrine of an originally integrated context, when man was not alienated from man or nature; or a fall, when private property was instituted; of salvation, when the revolution of the proletariat brings about the classless, stateless utopia through the birth of the new "social" man. The achievement of this utopia is arrived at through practical action.
>
> The principle of practice is given in a body of belief, dialectical materialsim, revealed by Marx. The citing of Marxian scripture as authority is proper usage. There are teachers and prophets, martyrs and saints of Marxism. In Russian Marxism there is a body of believers in communism, and a priesthood in the Communist Party and its structure. There are heresies - revisions, idealist and mechanist deviations. There is the demand for complete dedication to the point of successful Inquisition, as seen in the Moscow purges of the thirties. There is, as we well know, fervent missionary zeal. Further, as in any absolute system, there is paradox - the paradox, e.g., of free will and determinism...
>
> Even ritual is there, in the Party Congresses, in the celebration of Communist feast days, in the parades, in the "ikons" of Lenin, in the rite of visiting the tomb and relics of the first Russian communist saint. Near him now rises the statue of the first prophet - Karl Marx.
>
> Finally, there is a god, self-moving matter, making and acting in history...This matter is recognised as the ultimate and only power and reality. Thus it is worshipped.[48]...This affinity or parallel has been denied by Marxist officials.[49]

Struve[50] discusses certain pseudo-religious rites in the U.S.S.R. There are, apparently 22 "Miner's Days", "Railwayman's Days", etc. Special Communist ceremonies, deliberately anti-religious, were gradually introduced after 1920. There is some evidence to suggest that artificially conceived customs created by the bureaucracy are unlikely to flourish. For example, where the Russian carnival was psychologically necessary as a prelude to the asceticism of Lent, the transformation of the Carnival into a "Russian Winter Festival" has failed. One suggestion is that the old, traditional ceremonies be retained in a form suitable to Soviet society.

Harding and others [51] have analysed the Black Power movement and found both political and religious implications. There is, for example, a stress on "love". Black Messiahs and Marching Saints predominate, and especially after the death of Michael X there was a resurgence of the messianic theme. A constant theme is the Black Resurrection, when there will be new black men and a renewed, black orientated world. Both religion and politics intertwine, and the former is one of the major modes of expression for Black Nationalist movements.

H.W. Turner[52] is concerned in showing how the Christian missions and churches in Black Africa, in conjunction with the influx of Western culture, have been predominant in determining the growth of both nationalist sentiment and the achievement of political independence. He traces the rise of religio-political groups from the more violent movements such as those of the prophet Simon Kimbangu, the Mau Mau, and so on, to the less violent forms such as the Church of Christ, founded by James Limba in 1910, and the National Church of Nigeria, with its "God of Africa". The relationship between the political and religious components is infinitely complex and varies according to the historical period and nature of the group studied. What we can say with some certainty is that there exists a number of religious movements which have contributed to the development of political consciousness and action in Africa. Independent religious movements have united a tribal Africa to a surprising extent; there are even some reports of race being transcended. The search for a unifying ideology or religion to substantiate the endeavour for a united Africa has not yet materialised. Turner suggests that evidence necessitates us to "query the view that independent religious movements are no more than disguised forms of nationalism or of protest against colonialism; they have their own specific religious dynamic".[53]

What is apparent is that rising nationalist movements and leaders have utilised religious symbolism and ideology giving an impression of a structural unity between the independent church and the national political movement. This, however, does not imply that the new political leaders and parties themselves serve a pseudo-religious purpose, although it is sometimes tempting to assume this in messianic movements. What has, in fact, taken place is a transition from an ontocratic to a modern society; from a society which never demarcated sharply between a sacral and secular kingship or society to a society which does differentiate the two. Movements and regimes such as that of Nkrumah have messianic or religious features which are more superficial than real. When a relationship does exist between new independent religions and national secular movements it tends to be informal and loose, and any tie appears to be accidental rather than purposeful. As Turner states, the policy or formula is more often than not one of "render unto Caeser the things that are Caeser's and unto God the things that are God's". The new town of Aiyetoro, which was founded in 1947 by an Aldura group in Nigeria, seems to demonstrate this transition to a dichotomised secular/religious position.[54] What we fail to appreciate is

> the value of these indigenous grass-roots African movements that have so largely undermined the traditional mystiques of chiefship and ethnic group, established specialised societies for the purpose of religion, and allowed political development in independent Africa to proceed on an increasingly secular basis.[55]

Independent churches contribute to progress by collating both Western and traditional/local sources of culture, although some, especially the messianic movements, prefer a form of utopian escapism that contrasts sharply with a "this-worldly" social development.

Cohn's analysis of the Jehovah's Witnesses[56] parallels them with not only other radical religious sects but also the German youth movement <u>Wandervogel</u> and some of the Zionist youth groups, all of which he calls "proletarian". He also includes the Nazi and Communist movements. His use of the term proletarian is somewhat confusing, but basically he means by it any movement which lives in, but is not of, a society. Such movements are distinguished by a lack of participation in the society, by a sequence of separation rites, both of which serve to insulate them.

Despite differences there are also similarities in the proletarian movement. All emphasise some form of eschatology which "contains a particular esoteric knowledge system, a doctrine of rejection of the world, and a description of the millenium". All, too, lay some special claim to a certain body of knowledge, and to specific "facts" which are arrived at within a completely <u>closed</u> system of reasoning. The present order, whether it be capitalism or a satanic kingdom, is categorised as wicked and consequently rejected. Along with a rejection of the present world order, proletarian movements offer us a millenium; the Nazis the "Thousand Year Reich", the Jehovah's Witnesses "The New World".

The organisational structure is similar to all proletarian movements. These movements, while at the same time rejecting all formal organisational structure, also utilise it. All to some extent share a charismatic leadership. Often, as in the case of the Witnesses, there is no official leadership or position. The "onion" form of proletarian organisation means that the initiation process is one of a series of layers, the centre constituting the elite and the periphery the sympathisers. Coupled with all this is a deliberate attempt by these movements to introduce a process of mystification, often to the extent of actually inventing statistical figures.

Kiev,[57] in his analysis of the psychotherapeutic aspects of pentecostal sects among West Indian immigrants to England, points out that apart from the possibility of millenial ideas being compensations for the impoverished existence of sect members there may well be certain psychotherapeutic benefits accruing to them. The two prominent characteristic features of pentecostal movements or sects are a) patterned testimonials, and b) glossolalia. The gift of healing is also present, but it is directed more towards emotional and social repairing than physical defect and is called by the sect members "deliverance". Like witchcraft rites, failure is attributed to either lack of faith or an unsaved presence.[58]

> Although there is no evidence that emotional instability is a necessary ingredient for participation in the services, behavioural patterns institutionalised in the meetings are sufficiently broad as to provide suitable channels for the expression of a variety of needs and personality traits. For the depressed and guilt-written the sin-cathartic basis of the ideology and services provides a useful guilt-reducing device; for the hysteric a socially acceptable model for acting out; and for the obsessional the encouragement of a reduction of inhibitions and increased emotionality.[59]

A Durkeimian approach to social solidarity is presented in an intriguing article by Shils and Young[60] who maintained that society is held together by the common regard of the population to the sacred quality of certain fundamental moral standards which are funcitonally equivalent to religious beliefs, and commitment to them endows an overtly secular society with sacred values which unify that society into a single moral community.[61] The remoteness of the British Monarchy from the political scene is used to illustrate the way in which the affirmation of the Crown, through the ceremony of the Coronation, effectively substantiates the social structure of the society.

> In all societies, most of the adult members possess some moral standards and beliefs about which there is agreement. There is an ordering and assessment of actions and qualities according

> to a definite, although usually unspoken, conception of virtue. The general acceptance of this scale of values, even though vague and inarticulate, constitutes the general moral concensus of society.[62]

Birnbaum attacks the hypothesis on a number of grounds. The authors tend, in their analysis, to apply a framework which would be more appropriate to a simpler society. On the one hand they tend to think of the values as implicit and on the other as explicit. When they turn to their analysis of the society in which the Coronation ritual took place this seems as erroneous as their analysis of the Coronation ritual itself; there is too much factual emphasis on what they themselves term a symbolic ritual. In sum, "generosity", "freedom", etc., are too general to be applied just to Britain, and there is no concensus about these standards or values. Society does not exist at a purely unitary level but exhibits a plurality and multiple tensions. The Shils and Young thesis appears to Birnbaum to be motivated by an emotionally vested interest in such values as tradition, conformity and authority, rather than by a desire for analytical sociology.

Secular therapeutic parallels

Several secular organisations have been examined for their therapeutic value, and at the same time been shown to display covert patterns of transcendental structure and performance. While we are concerned with this as the general theme of this work it is nevertheless appropriate here to give some indication of the research in this field which reveals the general pattern of similarity between transcendental and non-transcendental groups.

An analysis of Alcoholics Anonymous[63] as a social organisation viewed it in relation to the particular ideological stance of the founder, and in particular to the recording of this ideology in the "Book" Alcoholics Anonymous. The groups themselves appear to display little articulated structure, and a significant gap exists between the ideological stance and the transference into an action system, although it is true to say that this is not the case in all A.A. groups. The practical running of such an organisation demands, in contrast to an ideology which stresses flexibility and lack of organisational formulae, an increasing emphasis on formulisation and rigidity.[64] "There are many organisational models which exercise a potential attractive influence on local groups - the model of some particular religious sect, perhaps, of some Lodge, service club, charitable organisation". Patterns of power, prestige, status and so on, are inevitable in a group structure of this sort, and so also are clique-patterns, conflict groups and other dissonant elements. Yet the overlying uniformity and solidarity rests on the shared experiences of alcoholism and the common aim of sobriety. The mechanisms of group confession and group therapy, although lay, do much to give these groups their structural identity and also to produce analogies with religious sects.

An analogous study[65] concerns itself with the self-help organisation, and specifically with Recovery, Inc. Although substantially fewer people belong to Recovery, Inc. than to A.A. (4,000 to 5,000 as compared to 350,000), nevertheless there are a number of elements in common to the two organisations. Recovery, Inc., was founded by a Dr. Low and the organisation's ideology is for the large part based on Low's book <u>Mental Health Through Will Training</u>. With the expansion of the organisation the personal influence of the founder decreased, and when Low died in 1954 the organisation became completely lay-run. Like A.A. it is non-sectarian, non-political, and is independent of any other organisation. The group numbers are, like A.A.'s, generally between 20 and 30. Recovery, Inc., like A.A., has a series of principles which provide an alternative cognitive framework for the members, and there is no formal criteria for membership other than having been mentally ill or "nervous". Meetings are now open to the public. The groups are more formally structured than A.A., and usually follow the pattern of a) reading from the Textbook, b) the presentation of examples (case histories), c) group participation, d) the question and answer period, e) the "mutual aid" social meeting. Wechsler makes several interesting points about the role of the Recovery, Inc. member. While it is a voluntary role, the patient has to recognise that he is different from the average person and yet similar to others within the group. Participation in Recovery, Inc. constantly reinforces the nervous patient role while at the same time providing no formal mechanism for terminating the role. The leader of the gorup also assumes a specialised role, as leader-expert, guardian of the method, and as a role model. Groups of this type will be examined in considerable detail in a subsequent chapter.

The treatment of the mentally ill can take many forms. They can be completely segregated in a psychiatric hospital, or they can be patients at a psychiatric day hospital, or, as part of the process of resocialisation, they can be attached to a "halfway house" such as Rutland Corner House[66] or a "sheltered workshop". Another alternative is for them to belong to an expatient organisation.[67] The organisation and structure of these latter organisations will be subsequently related to their ideological pattern. Such an attempt, to relate the structure and organisation to the ideology, was made by Rapoport[68] and Sugarman.[69]

An organisation which adopts a "hard" or tough line is one specialising in the treatment of drug addicts.

Sugarman[70] and Shelly[71] describe the organisation, structure and ideology of Daytop Lodge or Village. The newcomer at Daytop is received in silence and treated as though he is not there. The next stage in this initiation process is the interview where he is confronted by three men. They are, in fact, not social workers but ex-addicts, and after listening to his excuses and conversation for some time they begin the process of "devastation". "They continue to ridicule the now devastated addict before beginning a 'cleanup' operation." The only cause for addiction recognised at Daytop is stupidity.

> In other words, the concept of the drug addict as an ill person and therefore automatically entitled to the recognised perogatives of the role of the ill in our society in terms of sympathetic understanding, special concern, leniency and forgiveness is vehemently fought as an ideology. (in italics)[72]

The principal formal resocialisation agency is group therapy. There are no formal leaders but each group contains one person trained in specific forms of leadership. No psychological or sociological causes are accepted by the group who receive any such explanation with ridicule. The resident's behaviour is minutely criticised and the most vehement language is utilised. As Shelley says, the aim is to reach "gut level". The ideology is reinforced by a systematised status ladder whereby one moves up the rungs in relation to the length which one has been "clean". The Daytop philosophy is read aloud each day:

> We are here because there is no refuge, finally from ourselves. Until a man confronts himself in the eyes and hearts of his fellows, he is running. Until he suffers them to share his secret, he has no safety from it. Afraid to be known, he can know neither himself nor any other. He will be alone.
>
> Where else but in our common ground can we find such a mirror? Here, together, a man can at last appear clearly to himself, not as the giant of his dreams or the dwarf of his fears, but as a man, part of a whole with his share in its purposes. In this ground we can each take root and grow, not alone any more, as in death, but alive - a man among men.[73]

Secular aspects of transcendental groups

Pfautz's[74] work on Christian Science groups differentiates between the social structure of groups and the psychological texture (Heberle's phrase). It could be argued, of course, that the psychological texture must display itself in external behaviour, i.e., motivation is significant only as social behaviour, when it becomes action, and therefore it is a part of social structure. Pfautz argues that the changes which have occurred from its beginnings in about 1866 to the present day are to be described in terms of secularisation, i.e., the tendency "of sectarian religious movements and groups to become both part of and live in 'the world'".[75] Indeed, while the results were not completely in accord with the trend which he expected, there was some evidence that there was a decrease in affectual motives and an increase in purposeful-rational and traditional motives as defined by him. There was also an increasingly secular social psychological texture. The typology of orientations was based on Weber's schemata of orientations to social action - affectual, value-rational, traditional, and purposeful-rational.[76]

> Very generally, an affectual orientation or "motive" involes participation in a movement or group by virtue of some experience which arouses one's emotions. A value-rational motive is at stake when participation is a matter of a rational conviction concerning the goodness and validity of a group's goals. When the basis of participation is due to kinship ties, a traditional motive is postulated. Finally, purposeful-rational motivation...is proposed to be a matter of groups membership based on the expectation of personal advantage.[77]

In another paper Pfautz[78] argues that Christian Science has gradually acquired the characteristics of an institutional sect. He postulates five religious types representing increasing types of secularisation. These are the cult, the sect, the institutionalised sect, the church and the denomination. He bases his evidence on the demographic, ecological, associational, structural and social-psychological features of the movement.

Horton[79] discusses at length three principal attempts at defining religion. The first treats the term as being applicable to a wide range of human activity with no clear demarcatory lines. The second deals with "religion" as an oblique reference to social relationships and specific claims to social status. "The third type treats the term as referring to commerce with a specific class of objects, i.e., 'Religion is the belief in spirits' or 'Religion is the belief in the supernatural'."[80] Horton suggests that

> the way in which religious relationships are seen to differ from secular relationships in any given society is much the same as the way in which any one category of secular relationships is seen to differ from all other categories of such relationships.[81]

He argues for the assumption

> that in every situation commonly labelled religious we are dealing with action directed towards objects which are believed to respond in terms of certain categories - in our own culture those of purpose, intelligence and emotion - which are also the distinctive categories for the description of human action. The application of these categories leads us to say that such objects are "personified". The relationships between human beings and religious objects can be further defined as governed by certain ideas of patterning and obligation such as characterised relationships among human beings. In short, religion can be looked upon as an extension of the field of people's social relationships beyond the confines of purely human society. And for completeness' sake, we should perhaps add the rider that this extension must be one in which human beings involved see themselves in a dependent position vis-a-vis their non-human alters - a qualification necessary to exclude pets from the pantheon of Gods.[82]

If then one finds the variables used in an analysis of man-to-man relationships useful, then they will also be useful in an analysis of man-to-god relationships. The "treatment of a human population's relationships with its gods" is "a system of social relationships viewed in its manipulative aspect...". An important point by Horton is that every "major ontonological and epistemological category we can devise contains religious as well as secular entities".[83]

Mandelbaum[84] distinguishes between transcendental and pragmatic complexes in religion. "The transcendental complex is used to ensure the long-term welfare of society ...it is concerned with the ultimate purposes of man." On the other hand "The pragmatic complex...is used for local exigencies, for personal benefits, for individual welfare".[85] Whereas religion may at one time have been undifferentiated, societies tend today to distinguish, in varying degrees, between transcendental and pragmatic forms. These forms have their corresponding deities, rites, symbols, and practitioners, as Mandelbaum shows from his Indian sources. Also, transcendental and pragmatic systems coexist and mutually support each other. In advanced societies the very idea of "supernatural" has lost considerable ground.

> It remains to be seen whether the new secular beliefs and rites come to be functional equivalents for religion, even under the title of anti-religion. Spiro[86] has noted that scientific beliefs may serve as functional alternatives

> for religious beliefs, but that they cannot be functional equivalents since science cannot be invested with the emotional force of religion, nor can it serve the emotional needs of people as religion has done...
>
> If traditional forms of religion are taken lightly among some modern elite classes, the neo-transcendental forms bearing nationalistic or scientific labels are taken very seriously be many of them. Religious explanation...tends to be presented as firm and conclusive. The emerging transcendental doctrines are thus presented dogmatically, even though they are being revised continually. Just as there have been many revisions and competing formulations within the category of dual structure in religion, so may we expect a good many competing formulations within the ascendent category of transcendental religion.[87]

El-Islam[88] discusses the psychotherapeutic basis of Arab rituals. These rituals are transferred from generation to generation and acquire, by adherence to them anxiety-avoiding or anxiety-relieving properties. One such ritual is that of Sheikh Visiting as a form of religious therapy.[89] The Live Sheikh is consulted about various problems, and the practice is extended to dead Sheikhs and this "ritual is successful in people whose psychological needs to submit to a parent figure play a dynamic role in their character structure". Another ritual with a similar function is the Zaar, a ceremony consisting of rhythmic dancing and loud drumming, together with snake-handling.[90] Visitors participate to the extent of "collapse". Both rituals have a psychodynamic effect, and the fact that some benefit from them while others do not is to be explained by differing "object relationship systems during childhood and maturation".

La Barre[91] analysed primitive psychotherapeutic mechanisms in native American cultures. He was particularly concerned with the peyote cult and the existence of public confession which is evidenced by nearly all native Americans, and regarded it as a genuinely aboriginal psychotherapeutic techniqe.

> The great importance of public confession of sins in the liquidation of anxiety is obvious to the psychologist and the psychiatrist. The focusing upon some concrete fetish or symbol of cultural authority - the "Father Peyote", buffalo skull, spirit stone, or wampum - is another valuable aspect of the practice.[92]

The religious significance and influence on the concept of vocation or "calling" finds its most obvious example in the medical profession. Olsen and Whittaker's[93] account of the nursing world brings out a parallel between the nurses' Hospital Training School and a convent. Both are comparatively total institutions in which the novices are depersonalised and resocialised. Values and personality attributes are deliberately manipulated to designated ends. "Inevitably, in this secluded company, notions so much a part of both occupations arise: sisterhood, committed purpose, dedication of one's life to a 'calling', and sanctity of one's trust". Nursing also has heavy ritualistic parallels to religious orders.

Conclusion

Some of the secular parallels we have examined are too general, too inclusivist, to be of much value,[94] while others have indicated important areas where transcendental and non-transcendental groups or movements are analogous. Some groups, overly secular both in intention and practice, such as Alcoholics Anonymous and Recovery, Inc., appear at this stage to appropriate the symbol-forms of religious culture in order, as we have suggested, to facilitate or 'act out' their own group system. To accept the theory of functional equivalence is to postulate, as Wilson suggested, a naive perspective of a kind of unchanging "human nature", although on analysis there is some slight indication that certain secular forms of ideology have, to put it crudely, taken the "place" of transcendental ideological symbol systems. Similarly, certain non-transcendental groups and ideologies might, with reservation, be designated religious surrogates or pseudo-religions. What does seem to be certain is that belief systems or groups with ideological stances are occupying vaguely the same territory in the sociology of knowledge simply because they are legitimating systems. Collective ideologies are shared unificatory systems which utilise often common structural and organisational units.[95] We become pragmatically aware of ideologies in group manifestations of behaviour, and the internal structures of such groups have in common tasks such as goal-attainment, adaptation, integration, and pattern-maintenance. The boundaries of group ideologies have to be seemingly maintained, and input and output flows from the ideological system regulated and controlled.

A certain amount of overlapping both in organisation and structure is, therefore, inevitable. For example, reward structures, merit processes, authority structures, and so on, may often be the same in groups with manifestly differing goals. In other words, there is a minimum of constancy of interactive patterns and symbol forms that are necessary for the functioning of ideological group systems.

NOTES TO CHAPTER THREE

[1] This is the general idea conveyed by Guy E. Swanson, 1966, The Birth of the Gods, University of Michigan: Ann Arbor Paperbacks; and also by Philip E. Slater, 1966, Microcosm: Structural, Psychological and Religious Evolution in Groups, New York: Wiley. Both these will be examined in some detail later.

[2] Plato, Meno, 74 D-75A, translated by W.K.C. Guthrie, 1956, Baltimore, Md: Penguin Books.

[3] Ludwig Wittgenstein, Philosophical Investigations, 1953, translated by G.E.M. Anscombe, Oxford: Basil Blackwell.

[4] Ibid, Part I, section 66.

[5] William James, The Varieties of Religious Experience, 1902.

[6] Wittgenstein, op.cit., Part I, section 66.

[7] V.L. Olesen and E.W. Whittaker, The Silent Dialogue: a Study in the Social Psychology of Professional Socialisation, San Francisco: Jossey-Bass, 1968.

[8] Freud's model of the "mind" follows the platonic division of society into three parts - reason, craving or desire, will - and which constitutes the main thesis of The Republic. Freud divides the "mind" into the ego, superego and id. Freud literally thought of the "unconscious" as a place where motives were stored. This kind of theory, together with cartesian dualism, has no longer much, if any, validity with modern psychologists and philosophers. See Gilbert Ryle, The Concept of Mind, Hutchinson: London, 1949; Alan White, The Philosophy of Mind, Random House: New York, 1967; R.K. Jones, "Some Epistemological Considerations of Paradigm Shifts", Sociological Review, 25, No. 2, 1977, pp.253-271.

[9] Slater, op.cit., p.22.

[10] Swanson, op.cit., p.22.

[11] ibid., p.27.

[12] ibid, pp.9-10.

[13] ibid., p.42.

[14] T-groups are artificial structures and in one sense Swanson relied too heavily on such groups. The

"case of the absent leader" is confined to T-groups, and such groups generally have a high number of neurotics and middle-aged among their members. For a clear account of such groups see P.B. Evans, Improving Skills in Working with People: the T-Group, 1969, London: H.M.S.O., Training Information Paper 4.

[15] T-group conversation is stilted, artificial, and unreal. Slater quotes that of a person who referring to the leader says, "It was like being in front of an altar". Statements similar to this are utilised by Slater and others and interpreted as religious. See H. Ezriel, "A Psychoanalytic Approach to Group Treatment", British Journal of Medical Psychology, Vol. 23, 1950; W.R. Bion, "Group Dynamics: a Review", in M. Klein, P. Heimann, and R.E. Monet-Kyrle, editors, New Directions in Psychoanalysis, New York: Basic Books Inc., 1957; W.G. Bennis, "Defence Against 'Depressive Anxiety' in Groups: The Case of the Absent Leader", Merrill-Palmer Quarterly, Vol. 7, 1961.

[16] The theme of functional equivalence has appeared under various guises. See later in this chapter for Shils and Young's treatment of the Coronation as a functionally equivalent phenomenon.

[17] Bryan R. Wilson, Religion in Secular Society, London: C.A. Watts and Co. Ltd., 1966, p.xvi.

[18] It is not the intention here to give an account of the secularisation controversy. See David Martin, The Religious and the Secular, London: Routledge and Kegan Paul, 1969, especially the Bibliography on pages 157-161.

[19] Larry Shiner, "The Concept of Secularisation in Empirical Research", Journal for the Scientific Study of Religion, Vol. VI, No. 2, pp.207-220.

[20] Paul van Buren, The Secular Meaning of the Gospel, London: S.C.M. Press, 1963.
Thomas Altizer, The Gospel of Christian Atheism, London: Collins, 1967.
Harvey Cox, The Secular City, New York: MacMillan Paperbacks, 1965.
Gabriel Vahanian, The Death of God, New York: Braziller, 1961.
William Hamilton, The New Essence of Christianity, New York: Association Press, 1961.

It has been argued that of these only Van Buren, Hamilton and Altizer are atheists. See Colin Lyas, "On the Coherence of Christian Atheism", Philosophy, Vol. XLV, No. 171, 1970, pp.1-19.

[21] Thomas Altizer, op.cit. *en passent*.

[22] Amitai Etzioni, *Complex Organisations*, New York: Free Press of Glen., 1961.

[23] Liston Pope, *Millhands and Preachers*, New Haven: Yale Univ. Press, 1942.

[24] Bernard Barber, "Acculturation and Messianic Movements", 1941, *American Sociological Review*, Vol. VI, pp.663-669. Peter Worsley, *The Trumpet Shall Sound*, London: Macgibbon and Kee, 1968.
Vittorio Lanternari, *The Religions of the Oppressed*, New York: Mentor Books, 1965.
R.K. Jones and N. Gillespie, "The Saints of Ulster: Northern Ireland Paisleyism as a Millenial Movement", 1976. Unpublished.

[25] Peter Berger and Thomas Luckman, "The Sociology of Religion and Sociology of Knowledge", *Sociology and Social Research*, XLVII, No. 4, 1963.

[26] Ibid.

[27] B.W. Tuckman, "Developmental Sequence in Small Groups", *Psychological Bulletin*, 63, pp.384-399, 1965.

[28] C.G. Jung, *Psychological Types*, London, 1923.

[29] Werner Stark, *The Sociology of Religion I - Established Religion*, London: Routledge and Kegan Paul, 1966. Several sociologists have indicated a major weakness in Stark's major thesis in which he places the universal church *above* social influences. See Wilson, Bryan R. "Review article: establishment, sectarianism and partisanship", *Sociological Review*, Vol. 15, No. 2, July, 1967, pp.213-220.

[30] Ibid. p.137.

[31] Ibid.

[32] Ibid. p.138.

[33] Ibid. p.202. This is cited in Alfred Weber's *Schein und Wirklichkeit in der Volkswirtschaft*, Berlin, 1961, pp.129 and 130. Stark also points out (p.204) that the "scientific" Marx's indebtedness to the religious Felicite de Lamennais has not in fact been acknowledged by the Marxists.

[34] Ibid. p.206.

[35] Ibid. p.211.

[36] Ibid, p.213, Stark acknowledges his reliance here on F.C. Barghoorn, Soviet Russian Nationalism, New York, 1956. p.232 and 321.

[37] Ibid. p.223.

[38] Ibid. p.224.

[39] Ibid, p. 224, and mentioned by Stark as appearing in the magazine Esquire, October 1965, pp.129 and 159.

[40] Rudolph Heberle, Social Movements, New York: Appleton-Centure-Crofts Inc., 1951, pp.350-354. For a more sophisticated approach see John Wilson, Introduction to Social Movements, 1973, New York: Basic Books.
John Maynard, Russia in Flux, edited by S.H. Guest, New York: The Macmillan Company, 1942, pp.128, 293ff.
A. Rossi, in A Communist Party in Action, edited by W. Kendall, New Haven: Yale University Press, 1949.

[41] Heberle, op.cit., p.20.

[42] Maynard, op.cit., p.441.

[43] Alfred Stiernotte, "Mysticism and Communism", Journal for the Scientific Study of Religion, Vol. VI, 1967, pp.110-112.

[44] Gustav A. Wetter, Dialectical Materialism, New York: Frederick A. Praeger, 1963, pp.552-556.

[45] Alasdair MacIntyre, "The Christian-Communist Rapprochement: Some Sociological Notes and Questions", A Sociological Yearbook of Religion in Britian, 2, edited by David Martin, pp.173-186.

MacIntyre suggests that in order to have any contact both ideologies are forwarding eccentric claims and distorting parallels. He is not, I am sure, denying that some parallels of organisation and structure might exist.

[46] Mary-Barbara Zeldin, "The Religious Nature of Russian Marxism" Journal for the Scientific Study of Religion Vol. VIII, No. 1, Spring 1969, pp.100-111.

[47] For example, John MacMurray, Creative Society, New York: Association Press, 1936, pp.130-131, 126-7, and passim.
Nicholas Berdyaev, The Origin of Russian Communism, Ann Arbor: University of Michigan Press, 1960, p. 187.

Fulton Sheen, Communism and the Conscience of the West, Indianapolis and New York: The Bobs-Merrill Co., 1948, pp.109, 163, 168, 172.

[48] Zeldin, op.cit., p.109.

[49] 'Dialekticheski materialism' in Encyclopaedia of Philosophy Studies in Soviet Thought, III, 2 (June, 1963) pp.158-9. Signed: "T.B." (Thomas Blakeley).

[50] Nikita Struve, "Pseudo-Religious Rites in the U.S.S.R.", in Donald R. Cutler, The Religious Situation, 1968, Boston: Beacon Press, pp.757-764.

[51] Vincent Harding, "The Religion of Black Power", in Donald Cutler, op.cit., pp.3-38.
Essien-Udom, 1966, Black Nationalism: The Rise of the Black Muslims in the U.S.A., England: Pelican Books, pp.115-132, 183-194.

[52] H.W. Turner, "The Place of Independent Religious Movements in the Modernisation of Africa", Journal of Religion in Africa, 1969, Vol. 2, No. 1, pp.43-63.

[53] Ibid. p.48.
Peter Worsley, in The Trumpet Shall Sound, 1958, London: MacGibbon and Kee, seems to suggest rather the opposite, see pp.254-256.

[54] Turner, op.cit., p.52.

[55] Ibid. p.52.

[56] Werber Cohn, "Jehovah's Witnesses as a Proletarian Movement", American Scholar, Vol. 24, 1955, pp.281-298.

[57] Ari Kiev, "Psychotherapeutic Aspects of Pentecostal Sects Among West Indian Immigrants", British Journal of Sociology, Vol. 15, NO. 2, 1964, pp.129-138.

[58] When the sorcerer's spell fails to work he traditionally attributes this to mangu, to lack of faith or to the inefficient reiteration of the ritual. In this sense it is a completely closed system. If he attempts to "make rain" and this fails the explanation is simple: someone is present who is working against him in some way. See Max Marwick, editor, Witchcraft and Sorcery, London: Penguin Books, 1970, en passent.

[59] Kiev, op.cit., p.135. See also J. Spencer, "The Mental Health of Jehovah's Witnesses, British Journal of Psychiatry, 1975, 126, pp.556-9, and R.K. Jones (editor) Sickness and Sectarianism, London: Gower Publishing Company, 1984 (forthcoming).

[60] Edward Shils and Michael Young, "The Meaning of the Coronation", Sociological Review, Vol. I (II), 1953, pp.63-81.

[61] N. Birnbaum, "Monarchs and Sociologists: a Reply to Professor Shils and Mr. Young", Sociological Review, Vol. 3, 1955, pp.5-23.

[62] Shils and Young, op.cit., p.65.

[63] R.F. Bales, "The Therapeutic Role of Alcoholics Anonymous as Seen by a Sociologist", Quarterly Journal for Studies of Alcoholism, 28: 277-288, 1945. Both Alcoholics Anonymous and Recovery Inc. are elaborated upon in greater detail below.

[64] A similar point is stressed by Wilson in his account of sect organisation. See Bryan Wilson, editor, Patterns of Sectarianism, London: Heinemann, 1967, pp.14-15.

[65] Henry Wechsler, "The Self-Help Organisation in the Mental Health Field: Recovery Inc., A Case Study", Journal of Mental and Nervous Disease, 130, No. 4, 1960, pp.297-314. Also Wechsler, "The Expatient Organisation: a Survey", Journal of Social Issues, Vol. XVI, No. 2, 1960, pp.47-53.

[66] David Landy, "Rutland Corner House; Case Study of a Halfway House", Journal of Social Issues, Vol. XVI, No. 2, 1960, pp.27-32.

[67] Wechsler, op.cit.

[68] Robert N. Rapoport, Community as Doctor, 1960, London: Tavistock Publications. See especially, chapter 4.

[69] Barry Sugarman, "The Phoenix Unit: Alliance Against Mental Illness", New Society, 6th June, 1968, pp.830-833.

[70] Barry Sugarman, Daytop, unpublished MS, 1968. Also, "Daytop Village: A Drug Cure Co-operative", New Society, April 13th, 1967, pp.526-529.

[71] Joseph A. Shelly and Alexander Bassin, "Daytop Lodge A New Treatment for Drug Addicts", Corrective Psychiatry, XI (July, 1965), pp.186-195.

[72] Ibid. p.191.

[73] Quoted in Sugarman: 1968.

[74] Harold W. Pfautz, "The Sociology of Secularisation: Religious Groups", American Journal of Sociology,

61, 1955, pp.122-128.
Also, "Christian Science: A Case Study of the Social Psychological Spect of Secularisation", <u>Social Forces</u>, 34, 1956, pp.246-251.
Also Robertson, op.cit., pp.127-128.

[75] Pfautz, op.cit., 1956, p.246.

[76] <u>Max Weber: The Theory of Social and Economic Organisation</u>, edited and trans. by Talcott Parsons and A.M. Henderson. New York: O.U.P., 1947, pp.115-118.

[77] Pfautz, op.cit., 1956, pp.246-247.

[78] Pfautz, op.cit., 1955.

[79] Robin Horton, "A Definition of Religion and its Uses", <u>Journal of the Royal Anthropological Institute of Great Britain and Ireland</u>, Vol. 90, 1960, pp.201-226.

[80] Ibid. p.201.

[81] Ibid. p.212.

[82] Ibid. p.211.

[83] Ibid. p.205.

[84] David G. Mandelbaum, "Transcendental and Pragmatic Aspects of Religion", <u>American Anthropologist</u>, Vol. 68, 1966, pp.1174-1191.

[85] Ibid. p.1175.

[86] Melford E. Spiro, "Religion and the Irrational". In <u>Proceedings of the 1964 Annual Spring Meeting of the American Ethnological Society</u>, Seattle: University of Washington Press, 1964, pp.102-15.

[87] Mandelbaum, op.cit., p.1189.

[88] M.F. El-Islam, "The Psychotherapeutic Basis of Some Arab Rituals", <u>International Journal of Social Psychiatry</u>, Vol. XIII, NO. 4, Autumn, 1967, pp.265-268.

[89] W. Bazzoui and I. Al-Issa, "Psychiatry in Iraq", <u>British Journal of Psychiatry</u>, 1966, 112, p.827.

[90] For an excellent general description of similar manifestations in comparatively sophisticated society see, William Sargant, <u>Battle for the Mind</u>, London: Pan Books, 1959.

[91] Weston Le Barre, "Primitive Psychotherapy in Native American Cultures: Peyotism and Confession", Journal of Abnormal and Social Psychology, Vol. 42, 1947, pp.294-309.

[92] Ibid. p.307.

[93] Olesen and Whittaker, op.cit., pp.64-67.

[94] For example, the valueless inclusivist account in, Morris R. Cohen, The Faith of a Liberal, New York: Henry Holt and Co., 1945, pp.334-336, where he attempts, rather feebly, a comparison of baseball as a national religion in America.

[95] Berger and Luckman, op.cit.
Burkart Holzner, Reality Construction in Society, Cambridge, Mass: Schenkman Publishing Co. 1968, especially pp.143-162.

4 Sectarian characteristics of Alcoholics Anonymous

There is some suggestion that academics tend to distinguish rather too sharply between transcendental and non-transcendental belief systems. The distinction has certain heuristic advantages; to use the term "religion" to embrace some of the groups and movements which were mentioned in chapter three would embrace ideologies and groups which intended a separatist position. Humanist societies might be regarded as religious or sectarian groups for certain purposes of analysis, but clearly these groups signal a distinctive territory in the ideological sphere.[1] The confusion apparently lies with those who wish to make a logically erroneous jump from asserting that groups may have similar characteristics though public dissimilarities to concluding that similarity of characteristics argues for a unity of identity. The latter is essentially an inclusivist position as opposed to the former exclusivist one.[22] When pressed, however, sociologists at least concede that there is a continuum of belief systems from the strictly empirical to the completely metaphysical. We simply do not see the world in Otto's simplistic demarcation of sacred and secular. Not only do these elements often exist side by side, but there is some anthropological evidence that simpler peoples may make no such distinction.[3] What is perhaps sometimes obscured by the tendency to discuss belief systems as polar types, however, is the similarity of organisational and behavioural patterns that is found in certain secular and religious institutions. Thayer's popular description of fringe politics in Britain, for example, often echoes accounts of fringe religion:

> Those members whom I have met...are almost wholly those species of man who have sexual, psychological or physical axes to grind. Most of them have an abject contempt for the society in which they live. They are the misfits, the dullards

> and the outcasts of society who can think of nothing more constructive to do than to destroy all scions...[4]

Which is similar to McCluskey's description of the Witnesses:

> ...the bitter from whom life has stolen hope, the malcontent or anarchical who are eaten up with disgust at government, the business failures who have been bruised in competition...the simple, the illiterate, the dejected, the novelty seekers, the underprivileged...[5]

Fringe movements in politics are sometimes described as "sects". Essien-Udom's[6] analysis of Black Nationalism has shown it to be both a political and a religious movement. Susan Budd's[7] article of Humanist societies has brought out their sectarian elements. Again, Wilson, [8] following MacRae,[9] has shown an analogy between Marxism and Christadelphianism.

Alcoholics Anonymous is a movement which reveals distinct similarities to religious sectarianism both in organisation and belief. An examination of these may confirm or modify our conception of sect structure and that of value-orientated non-transcendental movements.

Alcoholics Anonymous

Alcoholics Anonymous, or A.A., is a world-wide organisation founded in America. Its World Directory of 1968 lists 8,484 groups in the U.S.A., 1,411 groups in Canada, and 2,784 groups in the rest of the world. With various prison and hospital groups the total number of groups is 13,000, with a membership of 350,000. Accuracy in these sorts of membership figures is obviously difficult to obtain, but the General Service Office[10] gives the estimated total membership as <u>more</u> than 350,000. The number of groups in the British Isles is approximately 250, with 2,400 members. These figures stand much the same today.

The founders of A.A. were Dr. Bob of Arkon, and Bill W., a former stockbroker of New York.[11] Both were approached by the Oxford Group (Moral Rearmament).[12] The beginnings of the A.A. movement were undoubtedly centred in religious experience. Early meetings had bible study, and there was some influence of the Dean of Bible at Wooster College who brought to the group's attention the restitution theme in the Sermon on the Mount. By 1937 relations with the Oxford Movement were severed. Among the early members were Joe, an upholsterer; Tom, a sheet-metal worker; J.D., a newspaper man; Bob, a "cookie" salesman; Phil, an insurance man; Barnie, Dick and Earl. Ritchie[13] in his survey finds the Twelve Steps[14] not unlike some of the "Washingtonians",[15] the Salvation Army, and the Oxford Group, these groups having advocated the following of certain of the principles long before the founding of A.A. There is some suggestion that it was the "Washingtonians" who first argued that the inebriate should be seen as a sick person, an idea which has been adopted wholeheartedly by the members of A.A., although Jellinek[16] gave it medical respectability. Ritchie maintained that the Oxford Group had the greatest influence and that A.A. took over the Group's ideas of self-analysis, restitution, the group confessional, group association, and the reliance on (some form of) God.[17] Undoubtedly the major influences have been group confession, surrender, guidance through the "quiet times", and sharing. A.A. Members hold, however, that the similarities are more apparent than real, as A.A. "is not allied with any sect, denomination, politics, organisation or institution".[18]

Organisational structure

The social organisation of social groups is concerned with the mutual interdependence and interaction of the constituent units of these groups. As an organisation there is a lack of rigid control and centralised authority. Local groups are essentially autonomous, and A.A. itself has as little organisation as possible. A group has an elected office of Secretary, and a large group a committee, both of which rotate. The "government principles" were first sketched by Bill W. when he founded the movement in 1935. The structure of A.A. comprises various units, and in America, in addition to local groups, we have the following pattern:

Alcoholic Foundation (Since 1957: The General Service Board of A.A.)	<u>Purpose</u>: administrating national affairs of A.A.
Central Office	<u>Purpose</u>: literature distribution letters, etc.
Works Publishing Company	<u>Purpose</u>: publishing A.A. literature
<u>Grapevine</u>	<u>Purpose</u>: newsletter
District Office	<u>Purpose</u>: arranging contacts, speakers
Control Committee	<u>Purpose</u>: advising
Clubs	<u>Purpose</u>: recreational
Auxiliaries	<u>Purpose</u>: anxiety alleviation for wives, etc.

A.A. organisation in Britain is at present in a state of change,[19] and no decision about the sort of shape it will ultimately reach has yet been arrived at. The Area and Group Representative Committee was the governing body for England and Wales and this committee appointed a General Purposes Sub-Committee which looked after the day to day organisation of the Fellowship. The Central Service Office was responsible for the Literature and the service to groups. The next step in the chain was the Inter-group, equivalent to the American District Office, which comprised a number of Area Groups, each Intergroup having a representative to the Area and Group Representative Committee. Scotland and Ireland adopted different structures so it has now been suggested that a Standing Committee be established to unify needs and policies.

In many respects A.A. has some of the characteristics of Etzioni's normative organisation: organisational orientation is marked by an intense degree of commitment,[20] and agreement rests primarily on internalisation of directives the acceptance of which is regarded as legitimate. The controls include rituals, social and prestige symbols, [21] resocialisation and leadership. Although there was undoubtedly some charismatic element in A.A.'s original organisation, it now has no distinct leadership of any kind. This contrasts with religious sects some of whom rely heavily on a charismatic role.[22] A.A. displays a democratic type social structure which rules out any necessity for external sanctions. On the whole this contrasts with religious organisations which rely, in the main, on normative powers to attain conformity to directives.

Both A.A. and sectarian affiliation to some extent share the same characteristics of voluntary associations. The theory of voluntary associations have ranged from the two-fold classification of expressive and social influence groups, the three-fold designation of instrumental, mixed and leisure, to degree of accessibility, status conferring and function. Other typologies have ranged from consummatory, instrumental and mixed designation, to behavioural, ideological, experiential, and consequential. Both A.A. and certain sects are instrumental-expressive, possessing low accessibility and low status.

Cultural and religious aspects of Alcoholics Anonymous

A.A. groups bear some resemblance to religious structure in their procedure. Most groups in Britain open with a minute of contemplation, followed by a talk, and ending with a self-confessional period. There is frequent audience-participation similar to that found in evangelical sects, frequent talk of God, and references to the "miracle" of sobriety. These procedures, of course, vary considerably from group to group. Members have a principle, which we might call "Nicodemus enumeration"[24]; "birthdays" are counted from the date of initiation at which time the member was "reborn". The organisation has adopted the prayer of the eighteenth century Friedrich Otinger: "God give me the detachment to accept those things I cannot alter; the courage to alter those things which I can alter; and the wisdom to distinguish the ones from the others".

The official handbook of A.A.[25] contains frequent religious intimations,[26] and such phrases as "Higher Power" occur often. Chapter IV is entitled "We agnostics". Often sudden revolutionary changes are described; there is a suggestion of "God-consciousness" followed by a vast change in feeling and outlook, not necessarily intended by A.A. The experiences are more of the "educational variety" described by William James, rather than sudden, because they develop over a long period of time. A favourite phrase of A.A. is "a Power greater than ourselves", although members are at pains to point out that this is not necessarily confined to a religious interpretation. Bales[27] suggests that there is a tendency for A.A. members to look back to the "founder" for guidance and to regard him with respect and reverence:

> Many religious sects and popular movements start in this way, and it is likely to be a major crisis in their existence when the key figure is removed, as eventually he must be, by death if for no other reason. If he has been able to transfer his "magic", or in other cases his "sacred" character, to a set of ideas, sentiments and procedures, perhaps expressed in a body of writings, or to some other impersonal source that can outlive him, the organisation has a chance to survive...Steps have already been taken to meet (this particular crisis in Alcoholics Anonymous). The "magic" has been transferred to "The Book", Alcoholics Anonymous, apparently with a considerable degree of success. One quite often hears of members who have become sober by the aid of The Book alone. The Book is the charter of the organisation, and the Twelve Steps are the core of its established procedure.

This to a large extent illuminates the particular role played by the editor of the Handbook. We know that in many sects lacking formal leadership roles, control of the movement came to those who managed publications; (it should be added that there is no evidence in A.A. that control was deliberately sought in this manner). Publications are important because they decide to a large extent the external image of a movement, while in a large organisation the printed word becomes more significant as influence by direct association becomes difficult.[28]

One of the obvious points of overlap between A.A. and religious sects is the dynamics of the therapeutics of fellowship. Because of the lack of formalised leadership an ideal or abstraction has been substituted for the concept of leadership itself; this we may call "metaphysical" leadership. The condition of membership for A.A. is simply "the desire to stop drinking" and there are no doctrinal requirements, members being free to believe what they will.[29] Nevertheless, both A.A. and sectarianism display certain characteristics in common. Stewart[30] suggests, as does Lindt,[31] that the initiate/sponsor relationship with its love-object/rescue-fantasy, plays an important part in the group membership of A.A. This is true also of many sects. Steward also suggests that as one of the characteristic tenets of A.A. is the acknowledgement of a superior power. Atheists or agnostics find it harder to recover than members who attend church.[32] This was partly borne out by a Merseyside study in which a majority of A.A. members found the content of the meetings too "religious" or "spiritual", although there is little doubt that this would vary from area to area and group to group. There is a religious factor in the dynamics of fellowship in A.A. which appears to be both useful and effective in the recovery process. A.A. "leaders" insist on their amateur status and never act as formal agents of control. Decisions are essentially group decisions, and as members of the same Fellowship there is no overt differential of status, although distinct patterns of deference do exhibit themselves in the interaction process. Self-confession to the group is as much a pattern of an A.A. group as of any evangelical or sectarian meeting.[33]

Lindt[34] showed that an A.A. therapy group conveys an atmosphere of euphoria just like any group experience, which can be seen as a direct parallel to the sectarian euphoria of Southern American groups as described by Sargent,[39] and Melanesian and Millennial sects as described by Worsley,[36] Barber,[37] and Lanternari.[38]

A word must be said here about the psychological process of conversion and surrender common to both A.A. and sectarianism.[39] This is a very obvious point where A.A. exhibits a similar characteristic pattern to sectarianism.[40] The changes which govern any particular shift are legion. Some alcoholics, like some religious converts, express a feeling of utter despair, while others experience some crisis of personal situation or behaviour. There is, a) the need to hit rock bottom, b) the need to be humble, c) the need to surrender, d) the need for the primary ego to be reduced.[14] There is very often a crisis moment preceding joining, which is in a sense the culmination of a process of strain. Some people appear to be under a process of regeneration. The A.A. group itself, like any sectarian group, is an unusually close primary community, which offers itself as a new reformist subculture. The conversion feature Tiebout says is "a psychological event in which there is a major shift in personality manifestation", a switch from negative to positive attitudes as follows:

formerly	presently
tensed	relaxed
aggressive	natural
demanding	realistic
conscience-ridden	conscience-free

and they occur in alcoholics generally when self-images of grandiosity and defiance are relegated. The conversion situation and its results are to some extent a positive therapeutic achievement, certainly in movements similar to A.A. On the other hand sect members often betray a hostile world attitude. Not all sects, of course, place an emphasis on an emotional shift, or a conversion experience. It is true, however, of more orthodox fundamentalist groups. It is what Ian Ramsey calls a "paradigm-shift", involving seeing things in a new light, and the adoption of alternative conceptual systems. The language used to describe it is almost identical with that used by scientists. It is a "gestalt switch", i.e., it must happen all at once.

To surrender, in Tiebout's usage,[42] is to listen and not fight back, and he describes an act of surrender when a patient attended an A.A. meeting:

> When I got there something happened - I don't know to this day...what it was, but when I took a look at the men and women there...I knew they had something I needed.

It may well be the case that A.A. should be considered a quasi-religious organisation: there would then be little difficulty in accounting for its religious features. The main objection to such a view is that A.A. categorically states that it is in no sense a religious body. Nevertheless, the structural features of A.A. procedure do resemble rather closely those of certain sectarian organisations.

Some sectarian characteristics of Alcoholics Anonymous

There is now quite general agreement regarding the characteristics of religious sects. Wilson[43] defines sect as "that small religious group in which membership is voluntary and conditional upon some mark of merit-understanding the group's teachings, or experience of some personal religious ecstasy - upon the basis of which association can arise". The sect is small and exclusive, and its voluntary membership is generally adult; authority is charismatic rather than hierarchically sanctioned. This is not to suggest that pure charisma is to be found in sects;[44] as in A.A. some sects display a poorly articulated leadership role, others a formerly instituted leadership role, neither of which need necessarily be charismatic.

Some characteristics are exclusively sectarian: separation from society in general and withdrawal from or defiance of the world and its institutions and values; exclusiveness both in attitude and in social structure; and attitude of ethical austerity bordering on asceticism. Nebuhr[45] saw the sect as a religious "conflict society" in opposition to an institutional church (society). Leadership also often has a charismatic quality. Not all sects display all these characteristics, and Wilson's definition seems to cover all manifestations of sectarian groups. A particularly important element of overlap between A.A. and sectarianism is the emphasis upon a "conversion"[46] prior to membership (the alcoholic must at least have the intention of relinquishing drink). Indeed, in the Merseyside groups 88 per cent admitted having experienced some form of "conversion". Other elements are: symbolism; voluntary joining; spirit of regeneration (the sect, like A.A., is an important "reforming" agency); single mindedness and fervour of adherents; exclusive bodies imposing some test of merit (the intention of abstinence in the case of A.A.); members kept under scrutiny and the pattern of their lives regulated in particular ways; sharing of central functions among the senior members; denial of any division of labour; equal total commitments; subjective fellowship and participation in mutual love; community and fraternity; the sect member both chooses and is chosen. A less certain element is membership by achievement: the longer the abstinence the higher the status within the A.A. group. Some sects, too, place an emphasis on achievement in the form of the recruitment of members. As previously mentioned, many members count their "birthday" from the time of initiation. Just as the sect tends to dominate a large part of the life of its members, to A.A. followers the group becomes a reference often for as many as seven nights a week. This is more than just a social pattern. The group accepts a totalitarian ideology which reorganises and reorientates the ideas, values and sentiments of its members. In a sense there is a parallel to be seen with the established sect, for the individual moulds and fits himself into an already existing way of life.

Some similarities of characteristics between A.A. and sectarian bodies have already been indicated. As formal organisations both A.A. and sects exhibit a set of purposeful rules and objectives which both prescribe and allocate certain obligations and status rights, and which to a large extent determine the activities of groups. Most sects and A.A. groups function with "the minimum degree of structure necessary to the successful functioning of the group as a social organisation".[47] Characteristics of a formal belief system that A.A. possesses along with many sectarian groups are: a) the nature of the organisation - structure loose and not strictly defined (although this is not true of Jehovah's Witnesses and Christian Scientists, for example), and the self-confessional aspect; b) teleological reference - a structured systems of goals, methods and taboos; c) criteria of formal inclusion.[48] The characteristics of formal organisation that A.A. possesses are: a) an unclear distinction along the formal/informal continuum. Indeed formal organisations often display informal characteristics, e.g., there need not necessarily be trained personnel in a formal organisation, and there are none in A.A. There is no formally recognised structure of organisation although unofficial patterns of power do exist; b) A.A. has little or no bureaucratic structure, the movement being independent of all other bodies, including other anti-alcoholic bodies,[49] although there is often co-operation; c) each A.A. group is an independent unity; d) A.A. has no distinct leadership and no particular allocation of roles. Generally, the longest standing member of the group acts as the chairman or speaker. Again, long standing is the determinant in the initiate/sponsor relationship; e) communication in the organisation structure is symmetrical, with roughly equivalent wholesale group participation; f) power and authority, also, is group centred, the group itself taking the role of decision maker.

Obviously, this is an extreme statement. There may be some sects that share the last five characteristics considered peculiar to A.A. Some sects may have a tight and well-defined organisational structure with no self-confessional aspect. Again, some sects may display patterms commonly thought to be confined to A.A., such as informal patterns of power as opposed to formally structured organisation, functionally dependent units, lack of distinct leadership roles, symmetrical communication patterns and group decision making. It will be apparent that there is some considerable interplay between A.A. and sect characteristics, and parallel organisational and behavioural patterns begin to emerge.

Religious value, alcohol consumption and sectarian affiliation

It is appropriate here to give some indication of research which has thrown light on religious, including sectarian, characteristics of alcoholics and A.A.[50] There has been a number of studies which relate alcohol consumption to religious beliefs and affiliation, and particularly to sect membership. Religious values have been shown as directly relevant to the recovery period in alcoholism. Some have implied that the religious content of A.A. is a vital contribution to its success. However, as we have already mentioned, there is some evidence that members find the tone of A.A. too religious. There is some suggestion in what follows of a connection between ascetic Protestantism and abstinence. This is especially manifested in the Protestant sects. If A.A. does have some of the characteristics of sects then part of its success may lie in this very fact.

Woodruff[51] examined the Christian conversion of the alcoholic in the context of the Protestant parish fellowship. He studied a group of twenty alcoholics who had been abstinent for a year and who had experienced some form of Christian conversion. They were aged from thirty-nine to sixty-two and predominantly middle or lower-middle class. He formulated four categories of conversion: a) psychosocial - involving improved interpersonal relations and entailing no religious interpretation; b) restrictive Christian - involving substitution of religious causes for compulsive drinking; c) limited Christian - in which the individual makes intellectual religious affirmation but no personal commitment; d) comprehensive Christian - total surrender of self to religious ideals.

Eckhardt[52] found from a study of case histories that the highest values for the recovery period were religion, friendship and generosity. Thorner[53] traced the relationship between ascetic Protestantism and alcoholism, and found a tendency for those with ascetic religious beliefs to be less tolerant towards drinking activity than those with Catholic beliefs, for example. The evidence he cites suggests that there is strong support for abstinence movements among the more ascetic Protestant sects, and indeed the nineteenth century temperance movements were originated by the Methodists, Congregationalists, Presbyterians and Baptists, all tending to be ascetic in emotional content. Thorner suggests that the Interpersonal Relationship system of A.A. sustains sobriety which is a primary goal. Parallels with ascetic Protestantism are as follows: union mystica, high premium on emotional discipline, and a call for help.

Riley and Marden[54] suggested a religious pattern of alcoholic drinking from an American National Sample:

Incidence and Frequency of Drinking by Religious Affiliation

	Abstainers (per cent)	Occasional Drinkers (per cent)	Regular Drinkers (per cent)
Protestant	41	46	13
Catholic	21	52	27
Jew	13	64	23

It is to be taken into consideration that drinking increases with urbanisation and that rural areas are predominantly Protestant. Myerson[55] found Mennonites, Quakers, Christian Scientists, and Unitarians as being particularly sober. Sessions[56] suggests that there is rebellion by alcoholics against religious orthodoxy, and a favourable response to "ego-religion" (individual conscience-centred) as opposed to "super-ego religion" (traditional authoritarian-centred). He further suggests a "spiritual" rather than a medical or religious approach to alcoholism.

Machoever, Puzzo and Plumeau[57] found no significant difference in the religious values of alcoholics as compared with non-alcoholics. Skolnick[58] found that 4 per cent of Jewish students experienced various kinds of social complications on account of drinking compared with 39 per cent of Episcopalian students, 50 per cent of Methodist students and 57 per cent of non-affiliates of abstinence background. That is, abstinence teachings, since they associate drinking with intemperance, inadvertently encourage intemperence in those who come from a background in which abstinence is stressed, but disregard the injunction not to drink. Frequently religious participation, even among students who drink, seems to diminish social complications.[59]

Several studies have associated religion with drinking patterns, Straus and Bacon[60] finding few Mormons were drinkers, and Tiebout[61] suggested that the religious content of the A.A. programme is the essential factor in the success of that organisation. Walters[62] studies the background of fifty alcoholics, 42 per cent of whom considered themselves Christian. More parents of the alcoholics were church members than parents of the control subjects. The controls were chosen from the neurological, surgical, medical and psychiatric wards of the same hospital, and matched for age and race. There was little difference between the two groups in early religious activity and church affiliation. More of the fathers of the alcoholics were heavy drinkers than fathers of the controls. Most of the alcoholics continue to hold their childhood beliefs about religion without any real change. In the Merseyside groups about the same number claimed they were Christian as in the Walters study.

Conclusions

It should be becoming abundantly apparent that characteristics of religious sects, religiously orientated movements, and so on, are not confined to those groups alone. The parallel or overlapping which is felt exists between transcendental and non-transcendental institutions, organisations, belief systems, etc., tends to be obscured, and it might be helpful if we regarded the two "polar types" as a two-directional continuum:

sect---------------secular associations

There is some suggestion that sectarian characteristics are to be found in various secular movements, and I have indicated that such sect-like characteristics are also to be found in organisations such as those which make no pretence at an overall ideological formulation. There is little doubt that closer examination would reveal similar characteristics in groups such as Neurotics Anonymous or Weight Watchers International. It is not the case that we can fit human behaviour neatly into categories which allow no overlap. Just as there are transcendental theistic or supernatural belief complexes which are, and are being, secularised (the Unitarians, for example) so there may well be secular complexes being transcendentalised.[63] The difference between the transcendental and non-transcendental to a large extent is in the interpretive patterning of human behaviour rather than in whatever overt or covert claim for legitimation of such groupings.

An evident criterion by which "secular" and "transcendental" voluntary associations may be connected is by their totality of concern. Alcoholics Anonymous, Daytop, etc., have the appearance of sects partly because of their total nature of involvement. Sectarians and A.A. members let these values determine other life stances and commitments. Denominations, on the other hand, conform more to the pattern of segmentary associations of a secular kind.

This chapter has examined the possibility that a particular secular organisation may have similarities to transcendental systems of a secular kind, and has attempted to analyse some of these similarities. A.A. in many respects resembles sectarian bodies of a certain sort both in structure and organisation, and this further suggests that irrespective of whatever claim to identification each particular organisation in fact makes, similarities can be discerned. Not only do secular movements often resemble, in structure, transcendental movements but secular movements have been known to develop into transcendental movements and vice versa. This is not to suggest that the identification is complete, nor that the distinction between the secular and transcendental is not a useful one, but in terms of sociological analysis we must accept that often those movements which appear dissimilar frequently possess an underlying similarity of structure. Difficulties remain. For example, the criterion of validation in a transcendental system is invariably super-empirical while a non-transcendental system formally rejects such validation. Manifest differences, may, however, conceal latent similarities, and similarities of structures and functions may camouflage differences in motivations and goals. The secularisation of a movement such as Unitarianism entails a redefinition of its terms of validation and legitimation. The Church, also, as a transcendental institution is increasingly involving itself in social and parochial work, as though the shift is becoming located in the empirical rather than super-empirical sphere.[64]

NOTES TO CHAPTER FOUR

[1] See, for example, Susan Budd, "The Humanist Societies: the Consequences of a Diffused Belief System", in Bryan R. Wilson (ed.), Patterns of Sectarianism, 1967, London: Heinemann, pp.377-406.
Colin Campbell, "Humanism in Britain: the Formation of a Secular Value-oriented Movement", Sociological Yearbook of Religion in Britain, 2. David Martin, London: S.C.M., pp.157-172, 1969.

[2] For an elaboration of this point see Roland Robertson, 1970, op.cit., pp.36-39.

[3] Robin Horton, op.cit.
Jack Goody, "Religion and Ritual: the Definitional Problem", British Journal of Sociology, 12, 1961, p.157.

[4] G. Thayer, The British Political Fringe, London: Anthony Blond, 1965, p.48. This is a description of members of the Union Movement.

[5] N.G. McCluskey, Who Are the Jehovah's Witnesses? New York, 1956, pp.15-16.

[6] E.U. Essien-Udom, Black Nationalis: the Rise of the Black Muslims in the U.S.A., England: Pelican Books, 1966.

[7] Susan Budd, op.cit.

[8] Bryan R. Wilson, Sects and Society, London: Heinemann, 1961.

[9] Donald G. MacRae, "The Bolshevik Ideology", The Cambridge Journal, 3, December, 1954, pp.164-177. This is a general, inclusionist, approach.

Wilson says (pp.352-353): "An interesting analogy occurs between the Christadelphian revolutionary hope and the communistic expectation of ultimate revolution. There are differences. Christadelphianism does not encourage men to involve themselves in direct social action; the Christadelphian is temporarily a quietist, and temporarily holds himself aloof from the world. But he is similarly disposed to regard the troubles of the world as the faults of the system; the wars which occur are likewise the wars of evil men and a base social organisation which in the more perfect dispensation of the post-revolution era (Marxist or Christadelphian) will not occur. Both systems promise relief to the dispossessed and discontented; both see the present social system as bringing its own downfall;

both are deterministic systems with conclusions already well-known to the initiates; both groups alone understand the secrets of the social process; both promise wrath to the unrighteous to be executed by the elect; both justify aggressive feelings in adherents by reference to noble ends." There are also differences between these two groups.

[10] This information was supplied by H. Toms, General Secretary, Central Service Office, London.

[11] The use of first or Christian names functions as a device intended to produce anonymity and enforce group cohesiveness.

[12] For a good sociological study of the Group see: A.W. Eister, Drawing-Room Conversion: a Sociological Account of the Oxford Group Movement. Durham, N.C.: Duke Univ. Press, 1950.

[13] Oscar N. Ritchie, "A Socio-historical Survey of Alcoholics Anonymous", Quarterly Journal of Studies of Alcohol, 9: 119-115, 1948.

[14] The Twelve Steps are basic to the A.A. programme of personal recovery. They made their first appearance in Alcoholics Anonymous, and are presumably based on the practical experiences of the early groups. A.A. members are encouraged to undertake a wholehearted approach to them. It was pointed out by Fr. Dowling, S.J., how A.A.'s Twelve Steps paralleled a part of the Exercises of St. Ignatius. They are to be distinguished from the Twelve Traditions which are primarily concerned with the groups as opposed to the individuals who comprise them.

[15] The extreme reluctance of A.A. to engage in public controversy stems from the example of the Washingtonians, who were a group of alcoholics who originated in Baltimore in about the 1850s. When their membership had passed the 500,000 the Washingtonians allowed themselves to be manipulated by politicians and reformers. For example, they divided on the issue of slavery.

[16] E.M. Jellinek, The Disease Concept of Alcoholism, New Haven, Conn.: United Printing Services, 1960.

[17] David J. Pittman, ed., Alcoholism, London: Harper and Row, 1967. en passent, but especially, Milton A. Maxwell, "Alcoholics Anonymous: An Interpretation", pp.211-222. Maxwell maintains that the third of the Twelve Steps, "Made a decision to turn our will and lives over to the care of God as we understood Him", has in practice enabled agnostics, etc., to embrace the therapeutic mechanisms of A.A. But

cf. David Steward, "The Dynamics of Fellowship as illustrated in Alcoholics Anonymous", Quart. J. Stud. Alc., 16: 251-262, 1955, p.255. "...many agnostics, even atheists, and nonchurch people in general, are not excluded from A.A. since the only requirement for membership is grounded in the common defect of alcoholism and in the common ideal of sobriety. These agnostics, atheists, and nonchurch people come to believe in the greater power, if they ever do, in the course of group dynamics by their own decisions. If they do not, it has been found that they find it harder to recover than those who do."

[18] A.A., 44 Questions and Answers about the Programme of Recovery from Alcoholism, London: A.A. Publishing Company, 1952.

[19] I am indebted to W. Kenyon, Executive Secretary, Merseyside Council on Alcoholism, for this information. (1973).

[20] Becker, Howard S., "Notes on the Concept of Commitment". American Journal of Sociology, LXVI, pp.32-30.

[21] One of the symbols used by some of the groups is an empty pint beer glass which is placed on the speaker's table during the meeting.

[22] See H. Wolpe, "A Critical Analysis of Some Aspects of Charisma", Sociological Review, 16, pp.305-318. I include here various African sects. It is true, of course, that leadership functions are often not charismatic, e.g., Jehovah's Witnesses, Christian Science, Christadelphians, Plymouth Brethren (all branches) Seventh Day Adventists, Mormons, have a thoroughly institutionalised charisma. Even among Quakers and Pentecostalists the charisma is considerably diffused. There are, however, some sects which are purely charismatic, e.g., the Father Divine Movement, the Lou-mensen, the Temoins due Christ Revenu. It is quite possible that both charisma and leadership are not attributes of individuals but arise from the relationship established between "claiment" and "believer".

[23] C.W. Gordon, and N. Babchuck, "A Typology of Voluntary Associations", American Sociological Review, 1959.
N. Babchuck and Alan Boothe, "Voluntary Association Membership: a Longitudinal Analysis", American Sociological Review, Vol. 34, No. 1, Feb. 1969, pp.31-45.
Kenneth Little, "The Role of African Associations in West African Urbanisation", American Anthropolo-

gist, Vol. 59, No. 4, 1957.

[24] St. John, Chapter II, verses 1-21.

[25] Anon, Alcoholics Anonymous, 1955, second edition, London: A.A. Publishing Co.
In 1940 the Alcoholic Foundation, founded in 1938, decided to take control of the Works Publishing, Inc. which was formed to publish Alcoholics Anonymous. By 1942 the Pounds 8,000 for paying off the shareholders was raised and the Foundation became the sole owner of the A.A. book.

[26] In the early days of the Book's writing there was some controversy regarding the content. The Liberals displayed no objection to the use of "God" throughout the book, but preferred "spiritual" to "religious". They refused to consider any other theological or doctrinal issues. The radical left wing were largely atheists and agnostics and consequently opposed to any religious or spiritual content whatsoever.

[27] R.F. Bales, "The Therapeutic Role of Alcoholics Anonymous as seen by a Sociologist", 1945, Quart J. Stud. Alc., 5: 268-278.

[28] I am particularly indebted to Dr. Bryan Wilson for this information. He pointed out that the editor of Christadelphian became in effect the leader of the movement - although not formally. In the Christian Science Church the directors took legal action in 1917 to gain control of the publishing house to consolidate their leadership. A similar story is to be seen with James Taylor, Jnr. in the Exclusive Brethren. The Jehovah's Witnesses place considerable emphasis upon publications such as Watchtower and Awake. Tudor, editor of The Morning Watch, the official organ of the Catholic Apostolic Church, or rather the Albury School of Prophets which preceded the sect, realising the importance of the publication, declined to transfer the office of editorship to anyone else, with the result that the paper itself ended in 1833.

[29] A step towards a more bureaucratised structure of organisation was undertaken in July, 1955, at the Twentieth Anniversary Convention of Alcoholics Anonymous, and expressed in the Conference Charter.

[30] D.A. Steward, 1955, "The Dynamics of Fellowship as illustrated in Alcoholics Anonymous", Quarterly Journal Study of Alcohol, 16: 251-262.

[31] H. Lindt, 1959, "The 'Rescue' Fantasy in Group Treatment of Alcoholics", International Journal of Group Psychotherapy, Vol. 9, pp.43-52.

[32] See note 17, above.

[33] This is particularly so in the case of evangelistic sects many of whom place considerable emphasis on the group-confessional. Note in passing the emphatic insistence of group confession during the Welsh religious revivals at the turn of the century. A.A. group confession derived much from Buchmanism.

[34] Lindt, op.cit.

[35] William Sargent, 1957, Battle for the Mind, London: William Heinemann, Ltd.

[36] Peter Worsley, The Trumpet Shall Sound, London: MacGibbon and Kee, 1957.

[37] B. Barber, "Acculturation and Messianic Movements", A.S.R., 1941.

[38] V. Lanternari, 1963, The Religions of the Oppressed, Vol. VI, New York: New American Library.

[39] "Nearly every A.A. has a spiritual experience that quite transforms his outlook and attitudes. Ordinarily, such occurrences are gradual and may take place over periods of months or even years.
A considerable number...who have had the sudden variety of spiritual experience see no great difference so far as the practical result is concerned between their quick illumination and the slower, more typical kinds of spiritual awakening." Alcoholics Anonymous Comes of Age, p.63.

[40] There are many points of similarity in the conversion processes which are common to value-orientated movements and beliefs.

[41] Harry Tiebout, 1949, "The Act of Surrender in the Therapeutic Process with Special Reference to Alcoholism", Quarterly Journal of Studies of Alcohol, 10, pp.48-58.

[42] Harry Tiebout, 1961, "Alcoholics Anonymous: an Experiment of Nature", Quart. J. Stud. Alc., 22, pp.52-68.
Some difficulties relating to seeing, believing, etc., are discussed in L. Wittgenstein, Philosophical Investigations, 1958, Oxford: Basil Blackwell, pp.192e-200e.

[43] Bryan R. Wilson, Sects and Society, London: Heineman.

[44] Some sects do exhibit characteristics of pure charisma, see above, note 22.

[45] Richard H. Neibuhr, 1929, The Social Sources of Denominationalism, New York: Holt.

[46] Not necessarily a "felt" change but more of a declared assent to sect values, e.g. teachings, beliefs.

[47] Bales, op.cit. This is not wholly true of sects, who often distribute "offices" not all of which are by any means essential.

[48] Budd, op.cit. p.382.

[49] The National Council on Alcoholism was begun in Britain sixteen years after A.A. was introduced here in 1947. The goals of the former are primarily two: a) educating the public about alcoholism, b) referring alcoholics or potential alcoholics to agencies. The emergence of the Regional Councils in Cardiff, Glasgow, Gloucester, Hereford, Leeds, Leicester and Stafford followed the formation of the first Council, the Merseyside Council on Alcoholism, which was established in Liverpool in 1963. Alcoholism Information Centres were located throughout the country. See G. Edward, M.K. Fisher, A. Hawker, and C. Hensman, "Clients of Alcoholism Information Centres", 1967, British Medical Journal, 4, pp.346-349, for a sociological description of clients frequenting such centres. Alcoholics Anonymous has formed Al-Anon family groups, and there are some 50 in this country and 300 throughout the world. The main purpose of Al-Anon is to "help the families of Alcoholics". Similar are the Al-Ateen groups.

[50] Benson Y. Landis, 1946, "A Survey of Official Church Statements on Alcoholic Beverages", Quarterly Journal of Studies of Alcohol, 4, 515-539.

[51] J. Woodruff, "The Christian Conversion of the Alcoholic in the Context of the Protestant Parish Fellowship". Doctoral Dissertation - Southern Baptist Theological Seminary: University Microfilsm No. 66-11264.

[52] W. Eckhardt, 1967, "Alcoholic Values and Alcoholics Anonymous", Quarterly Journal of Studies of Alcohol, 28, pp.277-288.

[53] Isoder Thorner, 1953, "Ascetic Protestantism and Alcoholism", Psychiatry, 16, pp.167-176.

[54] John W. Riley and Charles F. Marden, 1947, "The Social Patterns of Alcoholic Drinking", Quarterly Journal of Studies of Alcohol, 8, pp.265-273, p.270.

[55] Abraham Myerson, 1944, "The Treatment of Alcohol Addiction in Relation to the Prevention of Inebriety", Quarterly Journal of Studies of Alcohol, 5, pp.189-199.

[56] Percy M. Sessions, 1957, "Ego-Religion and Super-Ego Religion in Alcoholics", Quarterly Journal of Studies of Alcohol, 18, pp.121-125.

[57] Solomon Machover, C.S. Puzzo, and Francis E. Plumeau, 1962, "The Values of Alcoholics", Quarterly Journal of Studies of Alcohol, 23, pp.267-273.

[58] Jerome M. Skolnick, 1958, "Religion and Drinking", Quarterly Journal of Studies of Alcohol, 19, pp.452-470.

[59] In addition to Machover, Puzzo and Plumeau's work see: Norman C. Alexander Jr., and Ernest Q. Campbell, "Normative Milieux and Social Behaviours: Church Affiliations and Collegiate Drinking Patterns" in George L. Maddox, (ed.), The Domesticated Drug, 1970, pp.268-289, New Haven, Conn.: College and University Press.

[60] R. Strauss, and S.D. Bacon, 1953, Drinking in College, New Haven: Yale University Press.

[61] Harry Tiebout, 1946, "Psychological Factors Operating in Alcoholics Anonymous", in B. Glueck (ed.), Current Therapies of Personality Disorders, pp.154-165, London: Heinemann.

[62] O.S. Walters, 1957, "The Religious Background of Fifty Alcoholics", Quarterly Journal of Studies of Alcohol, 18, pp.405-416.

[63] The history of the Labour Churches, which were founded by the Unitarian minister John Trevor mostly between 1892 and 1895, demonstrate the difficulties which become evident when the secular and transcendental are married, at least artificially. Labour Churches were welcomed originally as a means towards a united Labour movement; religious interests tended to see them as either heretical or as clearing the way towards a more realistic form of religious commitment. The delicate balance between secular and religious was not, however, maintained, and none of the Labour Churches survived after the outbreak of the Great War. See K.S. Inglis, Churches and the Working Classes in Victorian England, 1963, London: Routledge and Kegan Paul, pp.215-249.

[64] The Churches have frequently, if not always, found it more politic to ignore socially adverse conditions rather than challenge the secular forces to implement

radical reform. Part of the fault lies in the ideological content of the dogmas of Christianity. From 1900 onwards, however, the Church of England in particular (Unitarians and Quakers, and Nonconformists had always displayed considerable interest and achievement in the field of social reform) awakened to the fact that its contact with the working classes was minimal and increased its efforts in the field of social work and social reform. It is as though in modern industrial society the religious sphere can legitimate its existence more by committing itself to secular reform than to passive reflection.

5 The organisation and structure of a transcendental group: the Catholic Apostolic Church

The Catholic Apostolic Church

People-processing organisations are concerned with the management of the identity of members in some way, either covert or overt, but these organisations differ in a number of respects, for example in the degree of the management which they exercise. Such organisations also differ in the way that the members perceive themselves, and the structure of the organisation, and different members will conceive the structure (or configurations of the "real" world as Nadel called it) differently simply because they are configurations. In the groups or organisations that we are examining it is clear that many of the characteristics described are peculiar to the groups in question, but although these are in themselves interesting our real interest is in the common dimensions exhibited by these organisations.

Our examination of Alcoholics Anonymous, a secular value-orientated organisation, revealed that it contained several characteristics of transcendentallly value-orientated organisations. Similarly, many other non-transcendental groups which were nevertheless value-orientated, such as the therapeutic movements, displayed characteristics remarkably similar to transcendentally value-orientated movements such as religious sects. In this section we will look at the origins, belief structure and organisation of a rapidly declining sect, the Catholic Apostolic Church, on the assumption that an account of this movement will reinforce the general thesis of organisational "overlapping" and shared characteristics peculiar to the movement which themselves have implications which are carried over to secular value-orientated movements, namely the nature of commitment and compliance. The second section will examine another sect, the Society of the New Church or Swedenborgians, in the hope that this also will give rise to a certain set of common dimensions.

Origin

The Catholic Apostolic Church[1] is historically associated with Edward Irving who was born in Annan, in Dumphrieshire, in 1792. His father was a tanner,[2] and his mother, Mary Lowther, was the daughter of a small landed proprietor. At school he displayed considerable ability and subsequently he fell under the influence of a Burgher Seceder named Hope with whom he used to travel to the small village of Ecclesfechan to attend worship at the kirk of the True Seceders. He graduated from the University of Edinburgh at the age of seventeen, in 1809, and promptly obtained the mastership of the school at Haddinton which he kept for two years. During the two years he studied for the ministry of the church and gave private lessons to a female pupil named Jane Baillie Welsh. Documentary evidence is lacking for this particular period but it appears that Irving fell in love with Jane Welsh who subsequently married Thomas Carlyle, another pupil at the Anna school who was later to achieve fame. In later life they remained friends of Irving. In 1812 he was appointed to a mastership at a newly built school at Kirkcaldy, eventually marrying a former pupil of the school, Isabella Martin, the daughter of the parish minister.

The religious climate of the Dumphries area of Annan was cold apart from a number of True Seceders,[3] but in 1815, the date Irving obtained his preaching licence, an Evangelical Revival began to sweep the kirk. Four years later he went as assistant to Dr. Chalmers, minister of St. John's parish church in Glasgow, who was later to become well known for his parochial system.[4] At this particular time Glasgow was almost on the brink of civil war which was being fanned by the revolutionary weavers and cotton-spinners, and undoubtedly Irving was considerably touched by the social conditions that prevailed.[5]

In 1822 Irving went as minister to the Caledonian Chapel or Asylum in London. The congregation numbered only some fifty members but Irving was tired of "the common stick of dry theology, the certified soundness of dull men".[6] His impressive physical appearance, his eloquence as a public speaker, and his undoubtedly "charismatic" personality, soon made him famous throughout London. A contemporary account recalled by Drummond describes the scene as follows:

> Sunday after Sunday the mean-looking, dingy chapel was thronged with statesmen, philosophers, poets, painters, and literary men; peers, merchants, and fashionable ladies were mingled with shopkeepers and mechanics, while many hundreds were unable to obtain admission.[7]

Many of the famous literary and political figures of the time, including Canning and Wilberforce, came to hear him. In 1823 he published An Argument for Judgement to Come and Orations, the latter being a protest against the visions of judgement of Southey and Byron. He was married in the October of that year to Isabella, having been unable to obtain a release from the betrothal. Jane Welsh Carlyle wrote years later: "If I had married Irving the tongues would never have been heard."

The apocalyptic fervour of the period, assisted by a general acknowledgement that the French Revolution was a portent of the impending end of the world, was contagious, and Irving's own apocalyptic period dates from 1826. Hair, in his book Regent Square, writes:

> If Irvingism is to be traced to its original germ, so far as any system can be traced to an individual, it may be found in Irving's religious experience, and in his consequent mode of apprehending divine truth not by open spiritual vision, but through a human vision."[8]

In 1826 Irving became familiar with the work of a Spanish Jesuit, Lacunza (pseudonym, Aben Eztra, sometimes Juan Josafet Ben-Ezra), which he translated under the title The Coming of the Messiah in Glory and Majesty. While attending a conference at Albury, in the same year, he met the second individual who was later to become prominent in the Catholic Apostolic Movement, Henry Drummond, M.P. The conference was to study unfulfilled prophecy.

In 1827 Irving moved to a new church in Regent Square; but Irving's star was falling, and the congregation soon settled at some 1,000 "sittings". In 1828 he published his Lectures on Baptism, which approximated to the theology of the sacramental party of the Church of England. In the same year a restless Irving journeyed to Scotland to proclaim the Second Advent. Oliphant records that he expected the Second Coming would occur in his lifetime, an event towards which he looked forward eagerly.[9] He was welcomed in Edinburgh, Carnwath and Bathgate. However, the death of 35 villagers in a throng at Kirkcaldy, due to the collapse of the gallery, cast a shadow over the tour.

At this time rumours of heterodoxy and controversy on the nature of Christ were directed at Irving. Though glorying in the Divinity of Christ, Irving held that the nature of the body of the Son of God was "one with us in all its infirmities and liabilities to temptation...we argue for an identity of life...". This view was attacked by a certain Mr. Cole, and by J.A. Haldane in his Refutation of Mr. Irving's Heretical Doctrine (1828). Irving replied in The Orthodox and Catholic Doctrine of Our Lord's Human Nature (1830) and Christ's Holiness in the Flesh

(1831). The concept of the holiness of Christ through the operation of the Holy Ghost led eventually to "Perfectionism" in the Irvingite Circle, and ultimately to the belief that "Gifted" persons were mouthpieces of the Holy Ghost when "in the power".[10]

> I believe that my Lord did come down and toil and sweat and travail, in exceeding great sorrow, in this man of temptation, with which I and all men are oppressed; did bring his Divine presence and death possessed humanity...and in that very state which God had put it after Adam had sinned, did suffer its sorrows and pains, its anguish, its darkness, vastness, disconsolateness and hiddeness from the countenance of God: and by His faith and patience did win for Himself the name of the Man of Sorrows and Brother and Finisher of our faith.

Events were building up against Irving. John McLeod Campbell, minister at Rhu, with his Doctrine of the Universal Love of God (that this Love was not confined to the Elect); Thomas Erskine of Linlathen; Hugh Maclean of Dreghorn; Scott of Woolwich; all were condemned for some form of heresy. A.J. Scott was Irving's assistant, and he had urged the latter that apostolic gifts - healing, prophecy, and glossolalia (speaking in tongues) - were given for all time and were lost only by lack of faith.

The presbytery of London, comprising six men, at least one of whom was personally hostile to Irving, challenged the heretical views. Whatever the verdict of the presbytery would have been, any such challenge was unthinkable in terms of the consequences. Finding no help in the National Scotch Church, Irving nevertheless insisted that at that time the Spiritual or Apostolic Gifts were being restored to the Church.

Glossolalia and other manifestations

The "unknown tongues"[12] were first heard on 28th March, 1830, from a certain Mary Campbell, a pious consumptive. Mary's sister Isabella of Fernicarry had died some time previously and was revered as a saint. Her minister Robert Story documented her life in his book Peace in Believing. The burn near to where she lived became a shrine. Near to Rosneath, in Port Glasgow, two illiterate brothers, the Macdonalds, came home from work and miraculously transformed their death-bed sister to health with the words "Arise and stand upright:.

Shortly after the miraculous events the Macdonalds and Mary Campbell began speaking with tongues which were attributed variously to the dialect of the Pelew Islanders, Turkish and Chinese. The account of the Macdonalds' tongues is recorded by a Robert Norton, M.D., in his Memoirs of

James and George Macdonald. Mary Campbell not only spoke with tongues but also exhibited automatic writing, and at Helensburgh people of all classes flocked to hear her. A similar crowd gathered at the house of the Macdonalds at Port Glasgow. The Albury members came from London. These members of the Prophetic Movement were all middle class, although they utilised the manifestations of the peasantry. It was concluded, inevitably, that the glossolalia were the voice of God, and that they were also the means of the evangelisation which was to prepare for the Second Coming. F.M. Davenport in 1906, in his book Primitive Traits in Religious Revivals, mentions similar outbreaks of glossolalia in certain Mormon communities which occurred almost simultaneously in the United States of America. In London a Miss Elizabeth Farncourt, the daughter of an Anglican clergyman, was miraculously healed. Two more instances of glossolalia followed, those of a Miss Cardale and a Mr. Taplin. These early manifestations were initially confined to small closed prayer groups, but in October 1831, the church in Regent Square was interrupted by an outbreak of glossolalia. Although Irving critically examined the increasing incidence of glossolalia, he concluded that they were indeed valid "signs", and that the original Pentecostal gifts were still manifest. The accounts of the manifestations in Drummond and Oliphant are psychological rather than sociological. The Albury circle was composed of nobility, lawyers, Members of Parliament, doctors, ministers of religion and teachers; nevertheless they welcomed the "signs". Irving viewed the Regent Square utterances as the work of the Holy Ghost, but others strongly denied this. Opposition to Irving grew, and he was finally charged with allowing unlicenced persons to interrupt public services. He was found unfit and deposed from his charge. His trial before the presbytery of Annan also resulted in a verdict of guilty. At the time the Annan verdict was declared Irving had left Regent Square church and had taken the majority of the congregation with him. He found temporary accommodation in a large bazaar in Gray's Inn Road, in premises which had been used by Robert Owen. In the Autumn of 1832 he moved again, this time to a picture gallery in Newman Street. On his return from Annan his Newman Street congregation refused him permission to exercise any priestly duties, although they granted him the position of deacon - the lowest order in the newly emerging hierarchy. A month later, in March, Drummond[13] related he was "called and ordained Angel or chief pastor" of the Newman Street congregation. He neither wrote nor appeared publicly again, and in 1834 he died. He himself had never exhibited any supernatural manifestations, and at the end he was forced to comply with those who had. He was buried at Glasgow Cathedral, and Oliphant records that crowds waited for a miracle at his tomb.

Of particular interest in the glossolalia outbreaks is the case of Robert Baxter, a High Anglican and a solicitor. He became a member of Regent Square and himself manifested utterances which he later came to doubt as being from God, and which he subsequently repudiated. These early manifestations were accepted genuine pentecostal manifestations by the early leaders of the Church, and many were later to become accepted as viewpoints.

The charismatic period

A new movement was beginning. Both the glossolalia and the Prophetic School were personal manifestations of quality without an institutionalised machinery. Between 1830 and 1833 there were no leaders or hierarchy or movement. The "routinisation" was to become manifest in a highly elaborate liturgy and office.

The Albury group, concerned with the study of prophecy, were organised and sophisticated, compared with the sporadic outbursts of the "tongues". It was at Albury that the liturgy and doctrines were formulated. It was from Albury, also, that the government of the Catholic Apostolic Church was initially exercised. The group first met in 1826 and annually after that for five years. Although concerned with prophetic writings they were also particularly anxious to receive guidance on the imminence of the Second Advent. Shaw, following C.W. Boase's Supplement to the Elijah Ministry, says they were agreed upon certain points:

1. "Premillennialism": that is, instead of the world becoming better gradually till the Millennium arrives, the present dispensation is to be destroyed.

2. The judgements are coming upon Christendom during which period the Jews will be restored to Palestine.

3. That the judgement will fall chiefly upon Christendom, the people of God according to their privileges being held responsible.

4. That at the termination of the judgement the Messiah will come.

5. That the Millennium will then be ushered in - a season of blessedness to all mankind and every creature.

6. That the vials of the Apocalypse began to be poured out at the French Revolution (1793), since the great period of 1260 years commenced in 533, when Justinian gave special recognition to the Papacy.[14]

An account of the proceedings is given in Dialogue on Prophecy. Henry Drummond, M.P., was the driving power of the Prophetic Movement. Two years after Irving's death he rushed to inform the Archbishop of Canterbury that the end of the world was near. Neither was he disheartened by the fiasco of prophesying and glossolalia. The Napoleonic wars hastened the feeling of imminence of the end, and Apocalypic fervour. Indeed, L.W. Scholler records some remarkable spiritual occurrences in Bavaria in 1827-28. Hatley Frere, a member of the Circle, had emphasised unfulfilled prophecy, especially in the Book of Daniel. It was to him that Irving had dedicated Babylon and Infidelity Foredoomed. According to Oliphant, Irving substituted for the millennial reign of Christ "the idea of a dispensation drawing towards its close, and...of an altogether glorious and overwhelming revolution yet to come, in which all the dead society, churches, kingdoms, fashions of this world, galvanically kept in motion until the end, should be finally destroyed".[15] Revelation and Daniel were utilised as symbolising the ending of the Reign of the Saints, which, as was asserted, lasted until 1793. To some extent the period following the Napoleonic Wars was one of economic and industrial unrest, and even people such as Arnold wrote:

> "I believe that 'the day of the Lord' is coming"

Timothy Dwight, President of Yale, often gave sermons on the imminence of the "Day of the Lord" which were influential.[16]

The official organ of the Albury meetings was The Morning Watch, edited by a Mr. Tudor, who later rose to eminence in the Catholic Apostolic Church. According to Oliphant [17] the early numbers were monopolised by Irving: "his name occurs, not so much an authority, as an all influencing, unquestionable presence..." Irving contributed articles on the Humanity of Christ, inquiries about the "spiritual gifts", and accounts of miraculous cures. His series Old Testament Prophecies Fulfilled in the New appeared also. In 1832 it was to condemn the General Assembly of the Church of Scotland which was about, like the "veil of the Temple", to be "rent in twain". It was in every respect also the organ of the original expelled community, controlled by the wealthy originator of the Albury band, Drummond. In 1833 The Morning Watch came to an end, having fulfilled its original intention of sustaining and perpetuating the spiritual utterances of the Albury School of Prophets, and introducing these utterances to the general public. The main subscribers were needed in other offices in what was emerging substantially as a new religious sect. Tudor himself became an elder of the new movement and declined to transfer the office of editorship to any other person.

One Nicholas Armstrong, a minister of the Anglican Church and later an Apostle of the Catholic Apostolic Church, preached: "Our inheritance shall be dominion over the creation; we shall rule angels: we shall rule the earth". [18] There was increasing stress on "signs" of the Second Coming. As Drummond says, "There was a craving", by the Albury School, "for the Gifts of the Spirit and an atmosphere conducive to charismata". The Prophetic Movement did not last long. The title Catholic Apostolic Church originated as a clerical error, but from these fragmentary, emotional beginnings, an institutionalised, hierarchically differentiated organisation rapidly developed. "This Church is Catholic in its use of ancient creeds and liturgy, based on Roman, Greek and Anglican rites; lights, incense and vestments are used. It is Apostolic in its permission of 'tongues'...and in its deacons, elders, prophets and apostles." The Second Coming was expected at the death of the last Apostle but when this time passed the movement abandoned any conversionist policy it might have had, either towards the Church authorities or laity.

The belief structure and organisation

The Catholic Apostolic Church believe in a restored apostolate, ecumenicalism, and an imminent Second Coming. The wanderings of the homelss congregation, headed by Irving, had precendent with the early Unitarians, many of whom, with the Toleration Act of 1689, left established churches and took their congregations with them. It maintained some allegiance to the Scottish Presbyterians, but gradually evolved a form of worship and a pattern of church life that were highly articulate and organised. As Whitley points out, this early congregation stressed two basic facts:

1. the permanent authority of the Bible;

2. the continued experience of the Church as the Body of Christ and the Temple of the Holy Ghost.

It was soon obvious that various streams were beginning to flow into it, among them Baptists, Anglicans, Congregationalist and Roman Catholics. It was from these early migratory congregations that the "Seven Churches of London" were formed; it was from them that the Twelve "Apostles" were drawn. There was a heterogeneity of origin unparalelled by any other sect.

There were two foci:

1. London was the "primitive" establishment;

2. Albury was the centralised headquarters for operations.

As we have already mentioned, it was at Albury that the development of worship and belief were formulated. There was the original Albury Prayer Book. By 1837 the first edition of <u>Offices and Liturgy of the Church</u>, prefaced by an exposition, appeared, though there were subsequent amended editions.

Several writers have pointed out that the architecture is ordered and set out for liturgical worship. Molloy,[19] in 1892, brings out clearly the architectural hierarchy.

At the top of the chancel steps were seven seats, the centre seat being for the Angel or bishop, the six others being for the elders. Below, in a parallel line, were seven other seats for prophets. Lower still, there were seven more seats for deacons, the centre one being occupied by the chief deacon. This arrangement represented "a threefold cord of a sevenfold ministry". The Angel ordered the service while the prophets spoke when the spirit moved them. The elders took it in turn to expound. The bread and the wine were given by the Angel to the elders, by the elders to the deacons, and by the deacons to the congregation.

After the death of Irving there was a rapid development of the movement. The full number of Apostles was called and a council was formed at Drummond's residence at Albury, on the model of the Jewish Tabernacle. Documents of intent were sent to the Archbishop of Canterbury, the Pope, and the Emperor of Austria. In 1840 the Apostles suspended the council.

"Holy elements" were always present during the services. The act of worship emerged as a highly formalised order of service, with the Eucharist eventually establishing itself as the central action, accompanied by an elaborate liturgical formulae. The liturgical sources include those of the Roman Catholics, Anglicans, Eastern Orthodox, and Lesser Eastern. In 1850 the Catholic Apostolic Church adopted the Reservation of the Sacrament. Before this, in 1838, there emerged a doctrine something akin to that of Baptismal Regeneration, and the setting up of altars demarcating areas of sacred territory. "Sealing" (the imparting of the Holy Ghost through the laying on of hands) was adopted in 1845, along with the introduction of chrism (consecrated oil). Elaborate vestments were introduced in 1842, altar lights and incense in 1852, and Holy Water in 1868. *The Liverpool Review*[20] for 1888 has the following:

> From recent statistics, the community appears to be making steady progress. It claims to have among its clergy many of the Roman, Anglican, Scottish and other churches, the order of these churches being recognised by it with the simple confirmation of an "Apostolic act".

Tithing, prayers for the dead and infant baptism were also introduced. In 1892, according to Malloy, there were 47 congregations in the United Kingdom, numbering 6,000 communicants. There were congregations in Ireland, Scotland, America, Germany, France, Canada and Switzerland.

The distinctive beliefs of the movement is the stress placed on the act of sealing, an act, like most of the ritual mentioned, which no longer takes place because the original apostles have died and cannot be replaced. This, we learn from Revelation 7: 2, 3, 4, had a distinctly adventist connotation, together with the stress placed on four of the Gifts of the Holy Ghost, i.e., tongues, miracles, healings and prophecy. The Church is elected from the world and is a divine society which is also holy and catholic and apostolic. It is a universal church exercising its function through the fourfold agencies of apostles, prophets, evangelists and pastors, and the corresponding fourfold ministeries.

The ministry of the Catholic Apostolic Church, under the second calling of the apostles, comprised the angel bishop or chief officer of the local congregation, the priest or presbyter, and the deacon. The subordinate office of underdeacon, now all that remains, is not a sacred order of the ministry, nor is that of deaconess.

In terms of eschatological commitment the Catholic Apostolic Church is adventist, believing firmly in the hopes of the early Christians. The millennium is stressed, and heavy reliance is placed on Daniel, Revelation, and the description of the Tabernacle (Heb. 8:5; Ex. 25: 9: 26: 30: 27: 8). Indeed, the whole order of the movement, some of the architecture of the buildings, is referred symbolically to the descriptions of the Tabernacle.

The Catholic Apostolic Church does not believe in a separate education of the young, and Sunday Schools are not advocated. Each location on the strata is fixed, and "the rich (will) be beloved by the poor, and the poor comforted by the rich". The movement is curiously uninvolved with social matters holding to a conservationist policy in relation to the social order and to the distribution of property. It is anti-socialist,[21] anti-temperance and anti-Salvation Army. It is ritualistic and authoritarian. There is a spiritual obligation to tithe.

The emergence of the sect

In the nineteenth century England was, contrary to preceding centuries and to the situation in Europe, characterised by a substantial period of peace which was broken only by the Boer and Crimean Wars. The influence of the changes which were occurring in Europe found expression in a reform of England's constitutional and political system, and simultaneously with an unprecedented rate of industrial expansion and invention. Notwithstanding, the abuses which were so much a part of the 18th century were perpetuated into the beginning of the 19th century. The French Revolution was, in fact, still making its impact at the time of the emergence of the Catholic Apostolic Church, and contributed to the period of 19th century English history the two great ideals of nationalism and liberalism. This era saw the emergence of secular orientated political theories such as communism, socialism and anarchism. Side by side with the growth of intellectual discovery there existed a singlemindedness which manifested itself in the military, political and cultural colonisation of much of the world. The attitude of the bishops to the events which culminated in the Reform Bill of 1832 made the Church of England as an established institution increasingly unpopular. The rapid social change that followed the major industrial advance perplexed and alienated the great migration into the urban areas, and shifted the balance of power from the landed aristocracy to the middle classes. The weak position and contribution of the Church of England, on its emergence into the 19th century, failed to counter the social conditions and the "new awakening" which sowed the seed for the evangelical and anglo-catholic revivals and the emergence of politically active Labour and Chartist groups.

To some extent the Catholic Apostolic Church pre-empted the subsequent movements of Tractarians and Plymouth Brethren. All were initially upper-class movements, and two at least were to some extent motivated by the frustration inherent in the clerical profession at this time. Upward mobility was virtually impossible[22] and the poverty of the clergy stood in stark contrast to the affluence of the church hierarchy.

Although often called "Irvingites" the movement denies this name. It originated through divine intervention rather than through the agency of any individual. Because it emphasised the manifestations Irving had to bow to those who displayed these manifestations. It adds that Irving's contribution was in the fostering of spiritual persons who were either driven out of their existing congregations because of the exercise of spiritual gifts or who saw that the movement in the form of Irving's original congregation was rich "in its numbers, ministries and gifts". Drummon describes him as "the forerunner and prophet of a new dispensation rather than the founder of a new sect".[23] Nevertheless, the movement inherited and still displays much of the influence of Irving. It denies, also, that it is a sect. Although professing they are not a sect, they share many of the beliefs which characterise sectarian groups. For example, they believe in the imminent Second Coming;[24] their interpretation of the scriptures is fundamentalist; they interpret political events, e.g., revolutions, as "signs of the times"; they are more willing to accept and emphasise prophecy than churches; they exhibit some intolerance of dissenters; they accept only the Nicene, Athanasian and Apostles Creed; they emphasise the doctrines of the permanence of the Gifts of the Apostles, Prophets, Evangelists and Pastors in the Church - the Pentecostal endowments are still with the Church; the supreme authority in the Church is exercised by twelve Apostles elected, not by men, but by God; the Fourfold Ministry, that of Apostles, Prophets, Evangelists and Pastors, and the Powers and Gifts of the Holy Ghost are to prepare and perfect the Church for the second Advent.

Denying their separatism, despite their separatist characteristics, they see the present period with only a handful of worshipping congregation as a sign that the Lord's coming is imminent. It is to the second Advent that they now look, and to the establishment of his Universal Kingdom of Righteousness and Peace. The hope of this imminence encourages them to penitence for sins which have marred his work, and to watch and pray for his appearing which is now nearer than they first believed. It is undeniably a separate gathering claiming a special core of knowledge. Most certainly it is not exclusive, and many of its members are also members of the Church of England. Its membership is voluntary and yet presumably its members gather together because to do so offers some distinctive reward. The number of children at a service is noticeable. Authority within the movement is both charismatic and hierarchically sanctioned. It possesses a formally instituted leadership role. It is not, in Niebuhr's sense, a "conflict society" in opposition to an institutional church. It is opposed to the idea of conversion in a salvationist sense and displays no fervour of adherence. Its membership, contrary to the assertion of Stark and others that all sects are proletarian in origin, is middle class both in its original founders and in its present composition. Its original apostles comprised an aristocrat, a member of Parliament, a solicitor, Anglican priests, an advocate of the Scottish Bar, a leader of the English Bar, an artist, a minister of the Scottish Church and a wealthy Northumberland landowner. The ministry of the Catholic Apostolic Church is purported to be prophesied by Malachi 4:5, and based on the construction of the Tabernacle. Each Apostle was given a portion of the world as his "tribe".

Between 1839 and 1840 Cardale, the Chief Apostle, was forced to recall the other Apostles from their work abroad in order to deal with a crisis of rebellion, which arose because some members regarded the council at home as the authority rather than the Apostles who were dispersed throughout the world. It was essentially[25] a dispute for authority between the Prophets and the Apostles. Drastic changes were introduced and the council was discontinued until 1847, and in fact never regained its original position.

By 1860 half the original Apostles were dead. Schism arose on the continent. The German Catholic Apostolic Church felt that more Apostles ought to be elected or called to bridge the gap, while Cardale in England stood firm in the conviction that twelve and twelve only had been called. The German faction based its policy on the claim that the number was exceeded many times in the New Testament (Acts 1: 26: 13: 2: 3: 14: 14, etc.). In 1865 it became The General Christian Mission and in 1908 the New Apostolic Church. It is a conversionist schism, working chiefly among Germans and those of German extraction. It is interesting that in Britain, Canada and America members of the movement were often able to carry out their obligations at the same time as holding offices or ordination in the English Anglican churches and the North American Episcopal churches, although the latter was less tolerant. In the Apostles Council in 1881 there were 31 members, representing Russia, France, Denmark, Germany, Britain and America.

Because they are the 144,000 mentioned in *Revelation*, members believe they are to be the rulers of the Church. They are diligent in stressing anonymity in their writings, and display complete uninvolvement with social problems. As Shaw says,

> ...the Catholic Apostolic Church is an emergency group having come into being in view of the imminence of the Second Advent...To all appearances the Church is dying out, and the people are content that it shall be so.[26]

The Catholic Apostolic Church was, in the beginning at least, the forerunner of later pentecostalist sects. The movement is unique because it combines a naive pentecostalism with a highly sophisticated ritual and liturgy. Irving's own theology was influenced not only by 16th century Scottish theology but also by his sympathy with the Prophetic Movement. The new sect, were it but known, was sowing the seeds of its later demise, by its emphasis on biblical literalism and the apostolic order of the Early Church.

Sect compliance is generally by normative commitment. However, as Wilson has pointed out, normative commitment is not a phenomenon which possesses unitary characteristics. Commitment is *positive* involvement, and positive direction with a high intensity.[27]

Involvement

	Direction	
Intensity	+	-
+	1	2
-	3	4

(Adapted from: Etzioni)[28]

Commitment varies not only in *degree* but also in *kind.* Wilson suggests that "Men may be primarily committed in different ways - emotionally, intellectually, ritualistically, morally or communally, to take some obvious respects".[29] Some sectarian commitments will emphasise one such way or ways, while others will emphasise other diverse commitments such as status, moral excellence, sobriety, etc. As Wilson points out, there may be various forms of commitment to religious movements which are neither religious nor normative. Often there is an open and relatively flexible framework of commitment, but "in the last analysis sectarian commitment is always voluntary".

Compliance and commitment

The members of the Catholic Apostolic Church have a diffused commitment for several reasons: historically, there has always been a close doctrinal, liturgical and membership tie, e.g., many of the original members were Anglican and subsequently it has not been in the least incongruous to hold membership in both the movement and the Church of England. The movement has never stressed exclusion or separateness, but rather has seen itself in the role of innovator, the vehicle of the apostolic age and the Holy Ghost. The Catholic Apostolic Church was called into existence by God. It possessed a highly formalised, professional, hierarchically differentiated ministry - the incumbents of this office often being simultaneiously holders of a priestly office in established churches. As a movement it does not fit happily into any of Wilson's types. It is actively opposed to a conversionist policy and to its concomitant doctrines. Although it is an adventist group it accepts the existence of the established churches, and it is not so much hostile as indifferent to the wider society. Although it has introversionist characteristics, it is not an introversionist sect; neither is it agnostic. In a sense it is occupying a uniquely marginal position in the church-sect typology.[30] On the other hand, its members do make special claim for their position in relation to the rest of Christendom.

Present position

Shaw quoted from the Manchester Guardian of the 12th December, 1934:

> The Catholic Apostolic Church now awaits dissolution within ten years, not because it has no funds - the tithe every member gives made it comparatively affluent - nor from the lack of devotees, but simply because the belief on which the Church was founded was that the Second Advent was at hand. Twelve apostles were appointed who alone had the power to ordain, and as the last Apostle died in 1901 all the surviving ministers are either elderly or aged. The youngest in London is sixty-eight. The one in charge of Paddington Catholic Apostolic Church is eighty-nine, and one who conducted services in Gordon Square last week - there are still services every day - was ninety. These men are carrying on because there is none to succeed them and it is realised the time is approaching when there will be none to do the work.

In 1980 the position was worse than the description of nearly fifty years ago. Some ten years later there were still some thirty-five churches in use. Shaw wrote in 1935:

> It is not likely that there are more than a few thousand of them all over the world. By marriage some are indeed admitted into the Group if so desired. Probably few people marry outside, those who do so being disposed to give up their Catholic Apostolic views.[31]

Today there is no church which offers a full service. The number of extant congregations, although probably not more than six in number, is not known for certain. The belief in the second Advent is stronger than ever, the decline in the congregation being read as a portent of the imminence of the events which herald that belief. The general patterns seem to be that worship is carried out by members of the Catholic Apostolic Church in the congregations of the established churches, principally, the Church of England, and adherents, unable as they are on the whole to meet for worship in their own churches, come together for meetings of prayer in each other's houses. The cycle of the movement has ordained that its end bears remarkable similarity to its beginnings, private prayer meetings being the way the movement started.

Conclusions

The circumstances which precipitated the emergence of a value-orientated movement such as the Catholic Apostolic Church and the period of emergence itself was one characterised by a surfeit of new movements, several value-orientated beliefs crystallized into actual movements in the 1820s.[32] Structural conduciveness was afforded by the preoccupation of the times with millennialism and prophecy, and religious ideology was not yet separated from the political sphere to the extent of value-orientated movements being excluded as conveying both secular and religious protest. Religion, being one of the dominant interests of society in the early nineteenth century, afforded an opportunity which was structurally conducive to the emergence of a religious value-orientated movement, but one which nevertheless afforded a camouflaged vehicle of protest for covert political anxieties. It was an attempt at a bourgeois revolution in Britain developed "under the guise of a religious reformation"[35] specifically precipated by the relative inflexibility of the governing aristocracy which was in effect blocking the social mobility of the bourgeoisie.

The emergence of the Catholic Apostolic Church at this time was due, if we follow the model afforded by Smelser, [34] to the impotence of the middle-classes in trying to restructure the social situation, in their inability to modify the normative structure, the facilities, or the mobilization components[35] available in the action systems.

The sources of strain within the social system might be located in the group's inability to cope with the rapid technological, constitutional and industrial changes which were occurring. This was a period of organizational strain, both economically and in terms of actual physical deprivation.

The social and revolutionary changes manifested themselves in a period of normative disorganization. These strains are "multiple and complex" and occur in clusters which "accumulate in different sequences...(and)...vary in strength and significance".[36] When some or all or more of these conditions of conduciveness and strain appear we get the crystallization of value-orientated beliefs which can be precipitated by such factors as give some indication or "reading" of a group's direction. For example, the glossolalia outbreaks, initially among the peasantry, were interpreted by the middle class Prophetic Movement as being manifestations of the voice of God, and also the means for the evangelization which was to prepare for the Second Coming. The movement's mobilization for action was facilitated by its flexibility regarding its own role and it saw itself principally as innovator and heralder of the apostolic age and Holy Ghost, and in relation to the latter it has never advocated a separatist policy. Once established it swiftly put into operation a formally instituted leadership role and established procedures for financing itself. The gradual demise of the movement was due to its belief in the imminence of the second advent, and not to its inability to adapt to existing institutions, nor to the disruptive effects of successive leaders or leading bodies, nor to the early schism which resulted in the succession of the continental branch. Originally a reformist and innovating value-orientated movement it increasingly adopted a negative and retreatist position. It was, in fact, a value-orientated movement which arose in a structural situation which was conducive to its initiation, and which also insulated, isolated and accommodated it.

The Catholic Apostolic Church exhibits a number of interesting characteristics. Firstly, its members display a diffuse commitment because of a long historical tie with the Church of England by both its officers and its congregations: it also perceives itself in the role of innovator. Secondly, it is a very apt example of a group that does not fit into a sect-typology but rather occupies a uniquely marginal position. This serves to remind us that commitment can take many forms and exhibit several degrees, and this also serves to remind us that typologies such as those that sociologists of religion are fond of displaying contain many human frailities. For example, they do not exist as "given" or "out there" but are analytical constructions of the sociologists. Very often they obscure the richness of group behaviour by typologising them with reference to only the more blatant and simplistic psycho-social characteristics of these groups.

Transcendental value-orientated groups such as this are formally organised socialisation settings concerned with identity management. Members may utilise the organisation itself as a mechanism of change but in the case of this organisation its decline is due to the very fact that its original ideology has undergone no change whatsoever. Voluntary affiliation in a group such as this is the result of habit rather than conscious recruitment. Members display a fine awareness of the rather complex setting presented by the Catholic Apostolic Church, e.g., the emphasis on elaborate liturgy, which in turn is reinforced by their tandem experiences in the Church of England. The members also bring with them into the sect's setting their own models of complex interpersonal settings. The result is a meshing of interpersonal setting models which enables the members to cope effectively with the management procedures utilised by the group.

Management groups such as this are primarily concerned with correcting deficiencies in the socialisation process or on a larger scale to deficiencies within the social system, and the organisational goals are manifested in specific goals contained in the socialisation agencies, e.g., the attainment of sobriety, remedying the sins of the world, and so on. All the groups we are concerned with are deviant in the sense that they are value-orientated movements concerned with behavioural aspects "out of tune" with those of society as a whole, that is, alcoholics, nonconventional beliefs whether political or religious, ex-mental patients, and so on, are merely examples of one <u>form</u> of organised social control exercised by society. The "coming together" into specific groups to share a specific ideology is the starting point for both a complex interactional setting and a coherent legitimated ideological standpoint.

The manner in which the Catholic Apostolic Church setting will differ from other resocialisation settings will be:

1. goals - whether the emphasis will be on <u>role</u> socialisation, e.g., educational training, or status socialisation;

2. recruits - whether the recruit content will or will not exhibit a high degree of variability in terms of age, sex, marital status, socio-economic position, and so on;

3. social interaction - whether this will be high or low within the organistion;

4. involvement - the degree or extent of the totality of involvement or commitment.[37]

Wheeler makes the point that the organisations that process people may differ widely in:

a) the degree to which there is a differential emphasis on or a clear separation between role socialisation and status socialisation. Bidwell,[38] has referred to this as a distinction...between...training and preparation for performance of specific tasks and...a broader pattern of training designed to prepare the recruit to occupy a generalised status in life with its associated lifestyles;

b) the extent of conflict that is felt between the needs of the recruits and theneeds of the external community;

c) the development of separate organisational hierarchies that give expression to these two differences; and

d) the extent to which recruits participate in the setting of goals and the adequacy of the fit between the goals of the recruits and the goals of the organisation.[39]

One would expect an organisation such as the Catholic Apostolic Church to exhibit a high emphasis on status socialisation, a low degree of conflict between the needs of recruits and the community, and a low degree of individual participation in the arranging of goal settings, with an adequate fit between the goals of individual and organisation. However, in some of the other groups, such as Recovery Inc., such a degree of coherence would not be expected nor is in fact manifested.

NOTES TO CHAPTER FIVE

[1] The standard works on the Catholic Apostolic Church and Irving are as follows:
M.O.W. Oliphant, The Life of Edward Irving, London: Hurst and Blacket, 5th edition, 1864.
Richard Garnett, article on Irving in Dictionary of National Biography, London: Elder Smith and Co., 1892.
Thomas Carlyle, Reminiscences, edited by C.E. Norten, London: Macmillan, 1887, sections on Irving and Jane Welsh Carlyle, Andrew Landale Drummon, Edward Irving and His Circle, London: James Clarke and Co. Ltd., 1937.
More recently, there is Professor Shaw's historical account of the movement, and Whiteley's small but significant introduction to Irving:
P.E. Shaw, The Catholic Apostolic Church, New York: King's Crown Press, 1946.
H.C. Whitley, Blinded Eagles: an introduction to the Life and Teaching of Edward Irving, London: S.C.M. Press, 1955.
Relevant correspondence can be found in the Clement Boase Collection in the British Museum.
See also the recent articles in the British Journal of Sociology by Wallis and Allan.
G. Allan, "A Theory of Millennialism: the Irvingite Movement as an Illustration", British Journal of Sociology, Vol. 25, No. 3, 1974, and R. Wallis, "Relative Deprivation and Social Movements: A Cautionary Note". British Journal of Sociology, Vol. 26, No. 3, 1975.

[2] See Stark Werner, The Universal Church, Sectarianism, for the important part played by tanners in the foundation of religious sects.

[3] Drummond, en passent.

[4] This parochial system was a microcosm of the Welfare State and included social workers.

[5] Oliphant, op.cit., p.53.

[6] Oliphant, op.cit., p.74.

[7] Drummond, op.cit., p.49.

[8] James Hair, Regent Square, London, 1889, p.42.

[9] Oliphant, op.cit., p.198.

[10] Drummond, p.113.

[11] Edward Irving, On the Orthodox and Catholic Doctrine of the Lord's Human Nature, London, 1830, p.2.

[12] Admirably documented by Drummond, pp.165-174. Also, Oliphant, pp.327-329. The Camisards, Fox, Joan of Arc, Swedenbourg, Southcott, the Shakers, the Mormons, etc., all manifested glossolalia at some stage.

[13] Drummond, op.cit., p.231.

[14] This also appears in a similar form in Edward Miller, The History and Doctrines of Irvingism, two volumes, London, 1878.

[15] Oliphant, op.cit., p.93.

[16] Stanley, Life, p.174. Dr. Bryan Wilson drew my attention to Dwight. See also: Latourette, Kenneth Scott, A History of Christianity, 1964, London: Eyre and Spottiswoode, p.1044.

[17] Oliphant, p.258.

[18] Drummond, op.cit., p.134.

[19] J.F. Molloy, The Catholic Apostolic Church, in Faiths of the Peoples, Vol.1, London: Ward and Downey, pp.74-101.

[20] Liverpool Review, June 9th, 1888, p.11.

[21] Shaw, op.cit., pp.219-223.

[22] Donald Capps, "John Henry Newman: a Study of Vocational Identity", Journal for the Scientific Study of Religion, Vol.9, No.1, Spring 1970, p.36.

[23] Drummond, op.cit., p.231.

[24] As a British millennial movement the hope of an imminent coming which never materialised at the time it was predicted but which is now felt to be very near bears similarity to the plight of the early Christians, the "poor" and the "saints" (Acts, 1:4; 2:46f) and also to such movements as Jehovah's Witnesses.
See: Joseph F. Zygmunt, "Prophetic Failure and Chialistic Identity: the Case of Jehovah's Witnesses", American Sociological Review, pp. 926-948, 1970.

[26] Shaw, op.cit. pp.234-235.

[27] B.R. Wilson, editor, Patterns of Sectarianism, p.7, London: Heinemann, 1967.

[28] A. Etzioni, A Comparative Analysis of Complex Organisations, Illionis: Free Press of Glencoe, 1961, p.9.

[29] Wilson, op.cit., p.7.

[30] Warburton makes this point in relation to the dual organisational commitment of members of the Faith Mission. "To many of its followers it offers a communal type of religion to complement the more associational nature of the denomination. It is in this very special respect that the Mission clearly lies outside the sect-denomination typology, although like the Quaker movement it maintains co-existent sectarian and denominational traits". T. Rennie Warburton, "The Faith Mission: a Study in Interdenominationalism", in A Sociological Yearbook of Religion, 2, edited by David Martin, London: S.C.M. Press, 1969, p.93.

[31] P.E. Shaw, Ph.D. Thesis, University of Edinburgh, The Catholic Apostolic Church, 1938, p.4.

[32] C.E. Sears, Days of Delusion, 1924, Boston: Houghton Mifflin, pp.xviii-xix.

[33] L. Trotsky, The History of the Russian Revolution, translated by Max Eastman, 1957, Ann Arbor: University of Michigan Press, Vol.1, p.14.

[34] Neil Smelser, Theory of Collective Behaviour, 1962, London: Routledge and Kegan Paul, p.325.

[35] Ibid.

[36] Ibid, p.342.

[37] Orville G. Brim, and Stanton Wheeler, (1966) Socialisation After Childhood, London: Wiley, pp.69-75.

[38] Charles Bidwell, "Pre-Adult Socialisation", Paper presented to SSRCC on Socialisation and Social Structure, May,1962, cited in Brim and Wheeler, p.70.

[39] Brim and Wheeler, op.cit. p.72.

6 The organisation and structure of a transcendental group: the Swedenborgians

The Swedenborgians

Both the Catholic Apostolic Church and the Swedenborgians or Society of the New Church are heralders of new insight and revelation. Both came into being as "restorative" movements - carriers of new truths. The pattern inevitably seems to be that such groups focus on a central figure such as Irving, Low, Swedenborg, Dr. Bob and Bill W., the early SPGB organisers, resulting in the formation of an organisation and distinctive "statements of belief". Typical of such groups is a reluctance to admit that they are, in fact, sects. Instead they are eager to place an emphasis on new truth, new insight and new knowledge, all of which sociologists claim as partial evidence of a sect-like nature. These groups also reject the taken-for-granted world. The Irvingites and the Swedenborgians reject or override the orthodox Christian tradition, the SPCB rejects other left-wing movements and has established its claim to being the true and only interpreter and vehicle of socialism, and Neurotics Nomine and Recovery, Inc., originally ancillary aids to professional therapy, are in fact offering a therapeutic milieu in the traditional pattern of mental health ideologies. They are, in fact, all "designs for living".

Such groups erect organisations which contain mechanisms for regulating derogatory significations of meaning which become, or may become, attached to the experiences of recruits. Commitment to such groups exhibits a totality in the sense not only of actual involvement but also because their sphere of influence extends into many other areas of members' existence. These groups not only magnify the importance of a central figure or new book but also lay special claim to be the recipients of new information drawn from a secret source external to the real world.

The following section traces the growth and origins of the movement known as the Society of the New Church and attempts to apply a social interactionist analysis both to that group and to subsequently raise general issues which might be applicable to the other groups, both secular and transcandental, with which we are concerned.

The most significant figure in the growth of the Society of the New Church, or the Swedenborgians[1] as they are sometimes called, was Emanuel Swedenborg who was born in 1688 in Sweden, the third of nine children in a pious Lutheran family.[2] "From my fourth to my tenth year I was constantly engaged in thoughts about God, salvation and the spiritual diseases of men, and several times I revealed things at which my father and mother wondered, saying that angels must be speaking through me." He was a man of wide interest and imaginative insight, and at various times he sketched proposals for a submarine, aeroplane, machine-gun and steam engine.[3] He received his doctorate at the age of twenty-one and travelled abroad, spending a year in London and Oxford in order to improve his mathematics, physics and natural history.

His first appointment was as "honourary Assessor" to the College of Mines which in 1724, when he was thirty-six, became confirmed as that of ordinary Assessor, together with a salary. For the next few years he devoted himself to scientific writing producing <u>The Principia</u> (1734) and <u>The Animal Kingdom</u> (1743),[4] among others, all of which were sustained attempts to create a coherent philosophical basis for the natural sciences of the day.

However, this was the final period of his scientific writing. In 1743, when Swedenborg was fifty-five, he claimed that he was subjected to a gradual revelation of spiritual insight.[5] The revelation came fully the following year: "(the) Lord Himself has appeared to me, and...he has sent me to do what I am now doing". He almost daily claimed contact with the Angel World. The spirits confirmed to him that drunkenness and homosexuality were "enormous sins", and these same spirits are said to have constantly harrassed him: "yea, also in the skin of my head, here and there with a slight hissing sound, like when some little distended visicle is perforated".

This turning point, of "amazing" gradual insights and spiritual conversion, made him withdraw from his work of Assessor in order to devote himself to the study of the Spirit World and the Old and New Testaments.[6] His programme of theological writing, which continued until his death, numbered thirty books, some, like his <u>Arcana Caelestia</u>, comprising several volumes.

Origins of the sect and schisms

In 1738, eleven years after Swedenborg's death in London, a certain Robert Hindmarsh gathered together a small group of five in his home in Clerkenwell in order to read the theological writings of Swedenborg. He had become aware of the work in 1778 when Swedenborg's Heaven and Hell was being printed by the firm for which he worked. Hindmarsh was the son of James, who later became the first of the new movement's ministers, relinquishing his post as writing master at the Methodist School, Kingswood School near Bristol, and ending his Wesleyan preaching. Robert was born in Alnwick, Northumberland, in 1759, and educated at his father's school. The small group which met in Robert's home developed into the first public meeting at the London Coffee House on Ludgate Hill. From there it adjourned to the Queen's Arms Tavern. It was but a short move to the establishment of a separate organisation which called itself the New Church, the original name having been "The Theosophical Society instituted for the purpose of promoting the Heavenly Doctrines of the New Jerusalem" (but in no way connected with Theosophy). The Great Eastgate Chapel was hired and opened for public worship on January 27th, 1788, with James preaching the first sermon. Robert himself was later ordained and was a leading figure in the New Church movement until his death in 1835, being President five times.

At the time of the Clerkenwell meetings - and others held in rooms at the Inner Temple and New Court, Middle Temple - the group had not yet separated itself from the orthodox churches and denominations. But early differences in interpretation of the writings of Swedenborg, on which the new sect based its beliefs, soon emerged. This process can be illustrated by reference to John Clowes, who in 1769 became rector of St. John's Church in Byrom Street, Manchester. Clowes became influenced by Swedenborg's Divine Humanity, and translated it from the Latin in an effort to influence others. He used his Anglican pulpit as a platform for the new teachings during the course of which he met opposition from members of his congregation. Hindmarsh wrote in the New Jerusalem Journal, 1792: "Readers of the new doctrine are very numerous in and about Manchester, many of whom have lately come to the resolution of separating from the Old Church and forming themselves into a distinct body, openly avowing and propogating the doctrines of the New Jerusalem". Clowes, however, was opposed to separation and believed that the new teachings would permeate eventually through the established churches. Although he made a special visit to London to plead his case to Hindmarsh and his friends the latter went ahead and formed a separate movement.

One of the earliest to come under the influence of Hindmarsh's public meetings was James Glen, a planter. Bound for New Guinea from England he broke his journey at Philadelphia, and in 1784 gave a lecture on Swedenborg's teachings. Swedenborg's works posted to Glen were later publicly auctioned and eventually led to the formation of a group of readers. The first American church was founded at Halifax, Nova Scotia, in 1791, and later groups were established at Boston and Harvard. This American development later became the General Convention of the New Church, a schism of the parent group in England.

The final schism occurred much later, in 1937, when there was controversy over the nature and interpretation of Swedenborg's theological writings. Over one hundred members resigned, chiefly in Holland, and formed themselves into a church body of their own which they now call "The Lord's New Church which is Nova Hierosolyma".

Order and liturgy of service

Molloy, writing in 1892, records that a service at Argyle Square, Kings Cross, began with a voluntary on the organ "during which the minister and his assistant, wearing white surplices extending to the feet and having voluminous sleeves, leave the vestry and take their places at the priedieus...a hymn is sung". The service continues in an orthodox manner with the first and second Lesson and Doxology, prayers and further hymns, and ending with the sermon, hymns and prayers.

The origin of public worship in 1788 introduced numerous liturgical styles to accommodate the new movement.[7] Some attempts were made to form approaches to worship which were uninhibited by these liturgies but in a very short time, that is before 1800, several examples existed. The best known early liturgy is that of 1802 printed by J. Hudson and Co., but with an unknown authorship and possessing no Conference (the legislative body of the Church) authority. The 1802 Liturgy bears the influence of the Book of Common Prayer but nevertheless was primarily important in introducing the Doxology into New Church worship. The 1820 Conference asked for "a form of worship which would be less objectionable than any now in use in those Societies which at present prefer extemporaneous worship". This led Hindmarsh and others to set up a committee in 1827 which in the following year produced a liturgy based on the draft of the form of worship used by Hindmarsh with the Salford Society. The "old" Liturgy made its appearance in 1875. One of the Committee members said that "In three months the firt edition of 3,000 was exhausted so I ordered 3,000 more, and then another 3,000". This 1875 liturgy came under a certain amount of criticism. "Conference, so often very wise, saw at once that any Liturgy, because it will be used as a vehicle to take men and women near to the Lord in worship, must be treated with great care in criticism, and showed clearly that it did not desire to hear more of negative comment."

The present liturgy which is in use made its first appearance in 1920, and is a revised version of the older one. Officially adopted by the Conference it is to all appearances a standard liturgical form containing the order of prayers before service, and the order of morning and evening services. The order of service for special days such as Ascension are set out, as well as the order of the sacraments for baptism, both child and adult "Holy Supper", marriage, burial, and specific prayers for the reception of junior and senior members, prayers for the sick, Te Deum, Psalms, and so on.

The organisation of the New Church

The first General Conference of the New Church (the legis--lative authority) met in 1789, although in 1821 it was reorganised as the General Conference of the Ministers and Other Members of the New Church signified by the New Jerusalem in the Apocalypse or Revelation of John.[8] The American Schism, after originally organising in Baltimore in 1791-2, adopted the name of "General Convention of the New Jerusalem in the United States" in 1817, although it was not until fifty-nine years later that a form of worship was adopted in America.

Further internal controversy occurred in the American group in 1890 when the Pennsylvania Association, which consisted of "fundamental Swedenborgians", insisted upon the virtual canonisation of Swedenborg's works and withdrew when a motion to this effect was not passed by the Convention. In 1897 this Pennsylvania group broke away altogether and became known as The General Church of the New Jerusalem with a presiding Bishop, a Cathedral and College at Bryn Athyn. The College is called the Academy, and publishes the Journal of Education.

The Conference of the New Church, the parent British body, is the principal and original organisation. The Convention of the New Church is the American group which is separately organised. The schism of the latter which resulted in the Bryn Athyn community in Pennsylvania, has itself a smaller division on the outskirts of this community which is also separately organised. The Bryn Athyn group has its own schools and colleges and virtual monopoly over the commercial and social life of the area. There are societies connected with these sections in other areas of North America and in Africa, Australia and Europe. In Britain there is a small day school at Colchester. The Conference itself no longer organises day schools. Affiliated to the General Conference of the New Church is the General Convention of the New Jerusalem in the United States of America and the Western Canadian Conference.

The sections of the movement differ over the emphasis and interpretation given to the theological writings of Swedenborg, such differences resulting in the forming of slightly different organisations. However, these sections maintain friendly contact with each other, and in 1970 an Assembly was held in London to celebrate the

200th anniversary of the publishing of Swedenborg's True Christian Religion, which was attended by members of all sections from all over the world.

The Swedenborg Society was founded on 26th February, 1810, chiefly by members of the Church of England, who having read Swedenborg's writings formed themselves into a non-sectarian organisation. It was incorporated in 1925, and is concerned with the promulgation, in definitive and scholarly form, of the works of Swedenborg, which have been published in English, European, Asiatic and African languages, as well as the original Latin. It also organises lectures and meetings to promote interest in the work of Swedenborg. The International Congress of the Swedenborg Society met in London in 1910, and consisted of 400 representatives from thirteen countries. Its counterpart in America is the Swedenborg Foundation. Both exist for the purpose of publishing Swedenborg's works which are the mainstay of the Church, and both are staffed entirely by members of the New Church or its sections.

In Britain in 1965 there were forty-eight societies or worshipping congregations with a combined membership of 2,933,[9] and fifteen study circles. In addition there were some 400 isolated members who are catered for by a publication called The Link. Other publications which appear regularly include The New Church Herald, and The New Church Magazine,[10] the latter being "more scholarly". The New Church also organises an extensive Missionary Society, Summer Schools, Correspondence Courses, Translating Panels for the Old and New Testaments, and the New Church College, the last existing for the training of ministers.[11] There are thirty ordained ministers of whom twenty-five are in active service, five serving overseas. There are six "ordaining" ministers, two of whom are still active. The Overseas Mission Committee superintends the New Church proselytising in Asia, South Africa (thirty thousand members), West Africa ("several thousands") and Europe, and the Continental Association of the New Church links organisations in Austria, Germany, Italy and Switzerland. There also exist other New Church organisations on the continent as well as auxiliary bodies and societies.

	General Conference (Great Britain)			General Convention (America)			
	1857	1900	1964	1850	1889	1899	1969
Members:	3,000	7,333	3,772	1,450	7,025	10,000	4,835
Societies:	48	76	56	54	-	-	65

The nature of the Swedenborg Revelation

The Society of the New Church established itself as a new sect, breaking away from the existing orthodoxies, and in turn itself experiencing schisms because of the importance attached to the "revelations" as manifested in the writings of Swedenborg, as follows:

1. The Second Coming has arrived in the form of the new writings, and there is no other authority or law.

2. The former Christian Church is dead. "Nor can there be a genuine Church except with those who separate themselves from it and come to the Lord in this New Church. The New Church is distinct from the Old, in faith and practice, in form and organisation, in religion and social life".

3. The Priesthood is the established means for the autonomous expression of the Church.

4. Baptism and Holy Supper and Conjugal Love are important components of the New Church.

The attitude to Swedenborg mainifested by his followers is such as to suggest that he is regarded not simply as a man but as a divine manifestation comparable to that of Jesus as viewed in Christian orthodoxy. The "Heavenly Doctrine" of the New Jerusalem as displayed by the writings of Swedenborg are "Divine and therefore infallible authority...(and one recognises)...these writings as being the Voice and the Word of God Himself, and not of a mere man".[12] In 1926 the General Church sustained this particular view by declaring "This Second Coming was effected by means of a man, His Servant Emanuel Swedenborg, before whom He manifest Himself in person, and whom he filled with His Spirit, to teach the doctrines of the New Church, through the Word, from him".[13]

The only serious contender to the paramount position of Swedenborg appeared in 1840, in a work entitled Diary Spiritual and Earthly of James Johnston, in New York. The original manuscript had been lost but a certain John Martin edited and subsequently published a diary version that he believed to be a valid facsimile of the original. Johnston records "Abraham met me according to promise, and amongst other things he said he and St. John would come on Wednesday and bring with them thirty-six kings". Johnston's Diary is in accordance with the diary records of Swedenborg but, notwithstanding, this particular contender made little impact on the new movement.

The doctrines of the New Church

The Word of God

The principle of Correspondence, a doctrinal interpretation of the Bible initiated by Swedenborg, states that the natural world corresponds exactly to the spiritual world. The Word has existed throughout time, and the Ancient Word was a source utilised by Moses to whom is ascribed the authorship of the Pentateuch. Before the Ancient Word there existed the Oral Tradition which accounts for the similarities in folk tradition drawn from many countries. The Mind of God corresponds to the universe which he has created. "...the six days of creation corresponding to six stages in man's reformation and regeneration. The Levitical sacrifices, prescribed for ancient Israel in such careful detail, became, under the key of correspondence, a revelation of the spiritual offerings of the repentant life."[14] The Letter of the Divine Word is limited to the books of Moses, Joshua, Judges, Samuel and Kings, Psalms and Prophets: four Gospels and Revelation.

The Second Coming

The Parousia is a spiritual event, as is also the Last Judgement. They are essentially past events which the New Church associates with the year 1757, when they believe the Last Judgement occurred at the "end of the age" (Matthew XXIV, 3), and that the Second Coming was effected through the instrumentality of Swedenborg. By this interpretation the Second Coming was into the world of Spirits because it was necessary that a human being from the physical world should observe and write about events in the Spirit world in order to convey his observations to mankind. The New Church believes that it was given to Swedenborg to be this observer and recorder, and that he wrote what he did without personal authority. "The Lord has manifested Himself to me His servant and sent me on this duty."

The Rejection of Tripersonalism

The dissenting bodies of the eighteenth century, many of whom were influenced by the Age of Reason, often made a rationalistic statement of doctrines and creeds. The variedness of the Protestant repudiation of Roman Catholicism contained many elements in common but can basically be divided into two main streams. The first was that which adhered to the Nicene and Apostle's creeds, the second was that which rejected these in an attempt to return to the primitive essence of Christianity. The rational climate set mankind firmly in a position to interpret the world around him. One of the dominant views of the eighteenth century was Deism which, proclaiming a natural religion, found no room for either the Incarnation or the Trinity. Many of the bishops and clergy in Britain were sympathetic to Unitarianism and Anti-Trinitarianism and Freemasonry too had some inclination towards Deism. Reason became the paramount instrument of both the interpretation of the knowledge of God and of man's total cognisance of Christian revelation.

It was in this context that the New Church rejected the view of the sophisticated doctrine of the Trinity established by the First Council of Nicaea in 325 A.D. for a theology of unipersonalism which members hold was precedented by Reason, the Old Testament and other great religions. It is in Jesus that the Divine Trinity of essentials, Father, Son and Holy Spirit, resides, and it is through Jesus's early life that the state of glorification is revealed to lead men to Redemption.

Heaven and Hell

These are states or conditions. Death is a form of status passage from one form to another. It is a change of experience. "Heaven is...a real world, a beautiful realm full of men and women angels." And those who are in heaven are continually advancing to the springtime of life: and the more thousands of years they live, the more delightful and happy is the spring to which they attain...In a word, in heaven to grow old is to grow young.[15] According to the New Church societies of angels vary, and some are concerned with the education of children while others are concerned with teaching those outside Christianity. Marriages are continued in heaven, if possible, and sexual intercourse is enjoyed not for the procreation of children but for goodness and truth.

The World of Spirits

The World of Spirits[16] is not the spiritual world, for the latter is the whole sphere of disembodied beings while the former is the area of those spirits who live neither in heaven nor hell but in between. In this respect it is similar to purgatory and the kamaloka of Annie Besant. It is a sort of antechamber that man enters into after death. Because it gives man the greatest possible freedom he is drawn to his "real" abode as manifested in the fullest expression of his own inmost character. Female angels, characterised by their love of children, greet and prepare any of the latter who might enter.

The nature of Swedenborg's writings and the New Church

Method of Biblical Interpretation

The New Church sees itself as the heralder of the new spiritual age of mankind, and it claims that its teachings have been revealed by God in the Writings of Swedenborg which contain a complete Christian theology. Considerable discussion has taken place by members of the New Church as to the exact nature of these Writings,[17] and their interpretation is important because of the implication for the sect's teachings. Basically the choice appears to be between regarding the Writings as Divine Revelation (Swedenborg was God's human, expositor instrument for interpreting the Word as contained in the Bible) or the Word itself ("dictated" by the Spirit of the Lord and exact to the very letter) which is limited to the portions of the Old and New Testaments mentioned above. Evidence from Swedenborg himself seems to suggest that the Writings were from the "Lord Himself" rather than from the angels:[18] that the complexity of the Apocalypse revelation from "the Lord"[19] has been granted only in this way to Swedenborg.[20] The implication from Swedenborg's own theological Writings is that he nowhere claims that his Writings are the Word - only the principles of the spiritual sense of the Word. The stress is on the newness of the manner of revelation.[21] Members differ on whether they believe the Writings to be the Divine Revelation or the Word itself. Swedenborgian fundamentalists claim that Swedenborg's Writings are the Word because of two sections to be found in New Jerusalem and Its Heavenly Doctrine (section 251) and Arcana Caelestia (section 10320): "What the Divine has revealed is the Word with us" (Quod Divinum revelavit est apud nos Verbum). "The Word is Divine as a whole and in every part, and the entire collections of sacred scripture were dictated by God."[22] As far as Swedenborg's own writings are concerned a commonly held view among members of the New Church is that the content of the writings were given to Swedenborg by "the Lord", nevertheless the choice of words was his own. "His Writings are truly the Divine Revelation of the Spiritual Sense of the Word but they are not the Word."[23]

Swedenborg's approach to biblical interpretation was in keeping with that of his age in regarding the testament as literal in the historical sense. In this he differs from an approach which utilises merely allegorical parallels to the testament accounts. Members of the New Church are not fundamentalist because they fully realise the contradictory accounts which derive from a literal interpretation. Spiritual Sense of the Word was the result of a non-literal interpretation contained within the natural world and summarised in the principle of correspondences. Examples of the Spiritual Senses exemplified in basic correspondence include the following:

> When the head is mentioned in the Word, then by the principle of correspondence, there is signified intelligence and wisdom. By the eyes is signified understanding; by the nostril, perception; by the breast, charity; by arms and hands, the power of truth...These correspondence of the body are built into the language. We say that an intelligent man has a head on him, that a perceptive man is quick-scented, that an intelligent man is sharp-sighted..."[24]

The nature of the Visions

Swedenborg's conversations with the Spirits occurred not only through the medium of dreams but also by

> changes in state when (he) was writing. Afterwards there were many visions when my eyes were shut; the light miraculously given; spirits influencing me as sensibly as if they touched my bodily senses; temptations also from the evil spirits, almost overwhelming me with horror; fiery lights; words spoken in early morning; and many similar events".

He was continually assailed by "flames of various sizes and different colours". His initial interest in the natural sciences was to enable him, so he subsequently claimed, to understand better the later revelations. His writings after his dramatic conversion were dictated to him by God. Some light might be thrown on his experience and condition by his longstanding ability to retain his breath for considerable periods without suffocating, an ability he seems to have shared with several mystics and seers in history. His initial vision and missionary institution were claimed by Swedenborg to have been given by God while he was living in London; and a person named Brockmer (living in Fetter Lane) recalls in William White's _Life of Swedenborg_ the foaming at the mouth and stammering that accompanied Swedenborg's experiences. "At last he said he had something to confide in me privately, namely that he was Messiah, that he was come to be crucified by the Jews." The outcome of the behaviour was that Brockmer became frightened for both his life and that of Swedenborg. "While I was with Dr. Smith (the physician) Mr. Swedenborg went to the Swedish Envoy, but was not admitted, it being post-day. Departing thence he pulled off his clothes and rolled himself in very deep mud in a gutter. Then he distributed money from his pockets among the crowd which had gathered." Although Brockmer met and talked to him on many occasions Swedenborg would never retract that he was Messiah.

An example of often bizarre visions included those of women, dogs and queens. "I seemed to be with Christ with whom I conversed without ceremony. He borrowed a little money from another, about five pounds. I was sorry that he did not borrow it from me." He appeared susceptible to visions even while eating his meals, and indeed his initial Commission was given in just these circumstances. He possessed a high sensitivity to bodily pains, and discomfort which pursued him during and after visions, and also an apparent ability as a seer which is well authenticated. Many tales developed around his ability to predict events in the future and to "see" events in the past.[25] Attempts were made to commit him to an asylum but to no avail. He died in London, where he spent his last days, in 1772, at the age of

84, having talked in his visions to Luther, Calvin, St. Paul, St. Augustine and Cicero.[26]

The background of the sect

The Society of the New Church bases the distinguishing characteristic of the movement on a rational approach to the "Revelation of God". One of the dominant theological ideas of the 18th century, Deism, was developed by such men as Edward Herbert and John Toland, who proclaimed a natural and universal religion, not mysterious but fully comprehensible by rational insight, and the Bible itself, formerly obscured in its meaning in many areas, was capable of clarification through this rational insight. The Swedenborgian movement emerged in an age in which the milieu was "rationalistic", and in which mankind was believed to possess an inner "Faculty" of "Reasons" or "Nature". The seed had been sown in theological circles by the "rational" theology of the Cambridge Platonists who emerged in opposition to the mechanistic materialism by Thomas Hobbes. The explanation of the hitherto mysterious resulted, according to the feeling of age, in a new knowledge or rational revelation, and those who adhered to this point of view were opposed to the emotional fanaticism and cant of many evangelical movements. Another source of influence, this time on Swedenborg himself, was most probably the Allegorists, and particularly the type of scriptural interpretation adopted by Sir Thomas Browne in the 17th century.

The Society of the New Church believes that the "Spirit of the Lord" revealed itself to Swedenborg, who is regarded as the divine instrument through whom such revelation was given with the instruction to convey to the human race what was from that moment "unveiled". It believes that his experiences of the "world beyond" were without parallel. By virtue of this special knowledge adherents, although maintaining that they are Christian, nevertheless place their emphasis on a new perspective of the old foundation, based on a rational and systematic interpretation of the scriptures by Swedenborg.

In a number of ways the teaching of the New Church differs from that of orthodox Christianity not only in its rejection of Tripersonalism or the belief that One God himself came to save mankind but also because it bases whatever doctrines it holds on the authority of the Word of God as defined by Swedenborg's explanatory doctrine on the grounds that the latter offer a model of rational faith. God's Will and Word are consequently believed to be revealed to men through the instrumental communication of Swedenborg's theological writings. The New Jerusalem and the life of angels, has for the sect's members, been graphically and immutably described by Swedenborg. Angels, for example, resemble humans in many ways except that their bodies are not material. However, they work, eat, copulate, and do all manner of things that humans on earth do. The central principle of New Church teaching is that of Divine Intervention by God himself.[27]

It is debatable whether Swedenborg envisaged the founding of a Church after his death.

> Since the Lord cannot manifest himself in person to the world, which has just been shown to be impossible, and yet has foretold that he would come and establish a new Church, which is the New Jerusalem, it follows that he will effect this by the instrumentality of a man who is able, not only to receive the doctrines of that Church in his understanding, but also to make them known by the press.

He claimed that true revelation had come from "the Lord" and not from the angels, and he appears to have attested to knowledge by revealed religion sustained on rational grounds. The New Church, it is believed, was established to restore what Christendom had lost, and act as a carrier of new truths. The emergence of this New Church suggests no disparagement on the existence of any other Christian church "so far as it is really the Lord's". It is, therefore, debatable whether Swedenborg wished himself to be seen as a founder of a new religious movement and members emphatically deny that he *was* such a founder.

In most respects the Society of the New Church appears uninfluenced by other pararelgious groups in the 18th century. However, some resemblances exist, for example the early spiritualist T.E. Spencer's communication with angels is similar to that of Swedenborg, and Brook Farm, a socialist community connected with the spiritualist movement, was Swedenborgian. Various scattered interests in magnetism (animal magnetism or mesmerism), especially in Sweden, appeared to have been influenced by Swedenborg's writings.[28] Indeed, although the Society of the New Church condemns spiritualism, it has been argued that the sect helped spread the growth of interest in postmortem communication. Theosophy, with its roots in India, founded by Madame Blavatsky and Colonel Olcott, and whose leadership eventually passed to Annie Besant, has some resemblance in the *kamakola*. The belief that humans have embodied the divine principle, for example Socrates and other sages, may have been influenced by the Swedenborgians. One of the first advocates of the New Thought Movement, Warren Felt Evans, was a New Church minister. Christian Science also has some parallels, particularly in the prominence given to Mary Baker Eddy's *Science and Health with Key to the Scriptures*, and in the avoidance of words such as "death" and "to die". In both groups members are referred to as having "passed on", "passed on to a higher life", and "passed into the spiritual world", and so on.

An interactionis analysis

The Society of the New Church originated during the last two decades of the Eighteenth Century, some sixty years before the development of the Catholic Apostolic Church. In some ways the two resemble each other. For example, both deny their sectarian character, and both were displaying at some period in their development strains of extreme reluctance to disengage from the mainstream of orthodox Christianity. Like the Catholic Apostolics, and like other sects, the Society of the New Church shares a variety of beliefs that marks it out as a transcendental value-orientated group. Members share a variance of a belief in a Second Coming albeit in a retrospective and spiritual sense. Adherents have their peculiar interpretation of the Bible and a concomitant regard for the Writings of Swedenborg. An emphasis on the World of Spirits further serves to denote the movement's sect-like nature. Its unique contribution, if we omit groups such as the Unitarians, is to rationalise the metaphysical.

Movements as vaired and dynamic as the ones we are concerned with cannot be easily limited to a structuralist-functionalist analysis.[29] They arise in, and are part of, a large society. Grandiose typologies which strive to mesh together stages or levels rarely, if ever, see these as interactive clements. Bittner[30] has suggested that radicalism as a term can be applied to any belief or movement which can be said to differ from ordinary, normal, traditionally sanctioned world-views which are current in any given society. In essence, a radical ideology <u>rejects</u> the taken-for-granted world-view. The common-sense view of the world about us necessitates the application of what Bittner terms "practical wisdom" in order for us to recognise its taken-for-granted nature, and the normative consequences of particular actions.

> A person recognised and treated as socially competent in a general sense, that is, a fully franchised member of his relevant community, is expected to move through such an ambiguous environment of objects and events with visible ease and to respect their existence as a condition of his conduct "until further notice". The stricture "until further notice" means that, given a special motive, the taken-for-granted and generally respected world-view can be made the object of a critical revision. Although critical revisions are somewhat out of the usual routine, one must not think of them as rare; they are the interstitial stuff in the web of daily routines.[31]

By applying Weber's discussion of prophecy to a secular area Bittner concludes that "by making the unity of meaning the defining characteristic of an outlook and way of life called radicalism, we indicate that an adherent adopts as morally binding an internally consistent scheme of interpretation and that he is, therefore, obliged to reason from a rigidly supreme principle to all occasions of actual conduct."[32] When the taken-for-granted world and its meanings is subjected to radical revision two processes can occur: a) the doctrine becomes impure - an example is the sect becoming dogmatised and bureaucratised into a church - this is what Weber terms the process of routinisation of charisma; b) impositions are introduced which show the believers that doctrinal impurity is both difficult and unattractive, usually through the process of "eliminating the possibility that believers will assign to any experience the significance of counter-evidence discrediting the professed faith".

The importance of Bittner's analysis to the main arguments of this work is that he supports the suggestion that the organisational task of the movement will consist in eliminating the possibility of derogatory significations of meaning being attached to believers' experiences. The three Parsonian mechanisms involved here are: a) gratification is comensurate with the movement's goals, i.e., members must receive a "pay-off" in some form; b) the internal structuring of roles, etc., within the group or movement must be so allocated as to minimise internal conflict and disharmony; c) a relatively rigid boundary must exist between the movement and the outside environment, and this can be achieved by restricting interaction with the external environment and by monopolising the interests of members. The organisational solutions of radical movements both secular and religious are seen by Bittner as follows:

1. A sense of Charisma must attach to the movement and its creed - and this Charisma can apply not only to the <u>person</u> but also sometimes to the <u>position</u> that he occupies.[33]

2. The doctrine that inspires the movement should contain information from outside the realm of everyday life.

3. On all rungs of membership there must be an intense concern for the purity of the movement's belief - this very purity becomes a means of control by the leader over the members.

4. No aspect of the life of a member can be defined as lying outside the scope of the doctrine or movement - what was described in the secion on Alcoholics Anonymous as <u>a totality of</u> concern. The commitment is total and all-embracing.

5. Suffering must become an integral part of a movement.

6. External sentiment should be exploited by the movement

for educational advantage. Obviously no radical movement, as Bittner states, will contain all of these characteristics.

All that has so far been said must be taken in the light of the specific situtational, interactive milieu in which groups arise and are perpetuated. Variables such as the constitutents of the groups, the distinctive doctrines and so on, will merely serve to make harder the task of applying a unifying set of characteristics common to all value-orientated ideological groups of this nature. If we return to Lauer's point,[34] these groups are part of an on-going interactive process, and his analysis, although concerned with Timothy Leary's LSD movement, follows Blumer's[35] identification of the distinguishing characteristics of such movements, "including a well-defined objective or goal, an organisation and structure, leadership, membership characterised by a 'we-consciousness', a guiding set of values, and normative expectations".[36] Movements of this nature involve, in their genesis, "an interactional process between an interest group and its social environment" and are characterised by an increase in stress, "a constriction of the legitimate avenues of conduciveness for the expression of activities of the interest group", and finally "a selective choice of alternatives that results in a sequential divergence of consensus between the interest group and its social environment regarding the legitimacy of the values, norms and/or goals of the former."[37] Consequently, these movements create symbol systems appropriate for the expression of the ideas or world-outlooks they contain, develop the idea of "messiah game", and the idea of social withdrawal. Previous attempts to account for the reasons why people become affiliated to value-orientated movements have ranged from explanations in terms of personal gain, emotional attraction to the leaders, believers in the cause, those with a stake in traditional impetus,[38] emotional or economic deprivation of some kind, for example Geschwender's attempt to explain adherence in terms of rising expectations, relative deprivation, downward mobility and status inconsistency.[39] However, the same argument was put forward very diligently by Stark[40] and the same refutations apply. Many movements such as LSD, the Catholic Apostolics and Society of the New Church cannot have their membership accounted for in any satisfactory way by a theory of deprivation. Lauer finally takes up Gerlach and Hine's argument[41] that the critical factor in adherence is face-to-face contact with members of this movement.

The question of what meaning a movement has to its adherents is answered by Lauer as a function of three factors:

1. the societal definition of the movement;

2. the adherent or potential joiner's own definition of the movement;

3. the movements' self-definition.

In the first instance, any movement or group of the sort that we are describing is regarded by society in a particular way, for example it might be regarded as "freaky" or "hippy" or "weird" or "occult", or it might be regarded as "strange" or even acceptable on a fringe level. The behaviour of the adherents may be such as to occasion no particular reaction in the external society because it falls within the bounds of normality or acceptability. Various combinations of factors enter into the total process, for example the movement, and here we come to the self-definition of the movements or groups, may not care what society thinks. On the other hand, it might take great care to present an appropriately acceptable image. The movement's self-definition is the way that it conceives its role, which will in turn determine largely the formulating of the ideology and vice versa. The adherent or potential adherent will, in turn, be a recipient of both the above, and his own definition of the movement will have to take into account both perspectives together with the situational factors from which he originated.

Before we apply the above approach to the Society of the New Church, and consider its implications for other value-orientated groups, it is worth mentioning Brown's analysis of the formation of ideology in a West Side Ghetto movement.[42] His concern was "to study that process by which poor negroes can recognise themselves as a communal entity, with common interests, a common field of action, and a common destiny".[43] Rather than being an informal gang the West Side Organisation for Full Employment has a more formal organisation, "with officers and programs, and with penetration of, and contacts with, the outer world of public officialdom, the mass media, the churches, and Civil Rights Movement". Basing his approach on Geertz's thesis of the concept of ideology in which Geertz argues that "ideology" as a term has become a "weapon word" for social analysis, Brown tries to demonstrate the validity of the former's contention that "ideology" as a term both names and evaluates what is named. Following from this the value of the study of ideologies, at least for Geertz and Brown, is that they are "indices of social perception, emotional orientation, and moral obligation". Consequently, Brown sees WSO as an attempt on the part of the indigenous negroes to enter the political area of public life and hence to reject the "received tradition" of American culture that the negro should exist in a non-participatory role in that area.

Finally, it remains to be demonstrated how ideologies

at variance with those in the "mainstream" of society can be sustained or maintained. Simmons' study of the Esper Community in Georgia[44] makes some important observations that can well be applied to other fringe groups and movements. Simmons lists five mechanisms which facilitate the maintenance of diverging ideologies:

1. Selective attention is paid to those perceptions which are congruent with one's own beliefs.

2. Social situations are actively structured so that their outcomes support one's own beliefs.

3. Ambiguous evidence is <u>interpreted</u> as confirming one's beliefs.

4. "Differential association and identification with those who share one's beliefs, coupled with relative isolation from and disparagement of those whose beliefs differ."

5. There is a marked ambivalence of the divergent larger culture towards one's beliefs.[45]

Because these groups and movements start off with a set of presuppositions or definitions of the situation this leads quite naturally to the process whereby a series of inferences can be made about the real world. Simmons argues that it is because of the ambiguity inherent in social situations that interpretation of the situation can be so construed as to be regarded as evidence of confirmation.

Salient aspects of social interaction in an analysis of the New Church

All the groups we are concerned with are subject to a sociological analysis which utilises perspectives such as fluidity, dynamism and variance. Because "reality is socially defined" such definitions are embodied in concrete and observable instances such as groups and movements in some form of social interaction. Furthermore, in order to understand how it is that a group or movement socially constructs the universe "at any given time, or its change over time, one must understand the social organisation that permits the definers to do their defining".[46] Such a method of analysis cannot be limited to a structural-functionalist perspective.

The movements we are discussing, inasmuch as they adopt as "morally" binding an internally consistent schema of interpretation, are in consequence "marked off" from the taken-for-granted world view. The Society of the New Church, like the other groups, can lend itself to an interactionist analysis which demonstrates a common ground for the organisation of experience arising from an interpretation of the everyday world. If we try to list a synopsis of the characteristics outlined here, with special reference to the New Church in this specific instance we can perhaps arrive at a general overall picture.

A. A charismatic element underlies the movement and its creed. This charisma can attach to persons (living or dead), positions, extraterrestoral knowledge, etc. Swedenborg is credited with special revelation.

B. Extra-terrestoral knowledge, in the way of religious or political beliefs, or even simply an emotional investment in a value-assumption about the effects of being an alcoholic, is implicit or explicit to the group's doctrine. The Writings of Swedenborg are a clear example of knowledge received from outside the realm of everyday knowledge or life.

C. All such movements embody a deep concern for the purity of their belief, and consequently various mechanisms of control are introduced or evolve. Schisms themselves become simultaneously examples of this purity and aids to a crystalisation of orthodoxy.

D. All such movements exhibit a totality of concern to some degree.

E. Suffering for "the Cause" - whether transcendental or secular - becomes an intrinsic aspect of such groups. The surrender of particular behaviour, whether over-eating, drinking, or "pursuits of the flesh" in general, is designated as purification through suffering.

F. Partly because of D such movements usually legislate either directly or indirectly the amount of external effective attachments which are allowed. Extreme examples are the marriage ties enjoined by the Strict Brethren.

G. External sentiment is often seized upon by the movement in order to achieve some educational advantage.

H. Objectives or goals are usually well-defined.

I. Organisations and structure are clearly recognisable, and so also is the leadership within these.

J. Aetiological growth stems generally from interaction between an interest gorup and its social milieu, and characterised by limitations of choice.

K. The meaning of a movement to its adherents is partly how society views or "tolerates" the movement.

L. It is also partly how the joiner himself defines the movement.

M. It is also how the movement sees itself.

N. An ideology of a movement is an index of social perception that offers a map of social reality which is the means of interpretation for the adherents to that movement.

O. Presuppositions in Collingwood's sense[47] underly all ideological assertions. They are statements about the world which are answers to questions stated or unstated, tacit or assumed.

P. Because they are attempts at imposing meaning upon the world such ideologies utilise mechanisms which facilitate the maintenance of diverging ideologies by enabling the adherent to "see only what he wants to see".

Q. Minority ideologies, whether secular or transcendental, unless they exhibit a marked variance with the beliefs indemic in the wider society, are treated with a marked tolerance and ambivalence in a pluralistic society.

The Society of the New Church is characterised by many of the above. Swedenborg is credited with a charismatic revelation, and his Writings are regarded, to a greater or lesser degree, as containing a form of extra-terrestrial knowledge. The movement itself is concerned with the maintenance of a purity of belief, and the American and Dutch schisms have merely served to crystalise the distinctive beliefs of the Swedenborgians. Members, in common with all the groups that we have examined, exhibit a totality of concern together with some form of voluntary deprivation for their beliefs. Behavioural patterns demonstrate that the movement's members, together with other movements and groups in society, whether secular or transcendental, exhibit concern and positive discrimination in social contact. Members of Alcoholics Anonymous, for example, will mix far more with each other than with "normal" society members. The Catholic Apostolic Church members and the Swedenborgians will tend to limit their external affective attachments to their own groups. All these groups exhibit a clearly defined organisation and structure. Quite obviously these groups are regarded as having quite separate meanings for their members than they apparently do for the rest of society. They tend to be very small and distinctive and their ideologies contain clear statements of how they "map" the real world.

In some senses the groups at which we are looking are at variance with the prevailing culture. Nevertheless, they are accommodated by the prevailing culture because they offer no real threat. To the adherents their ideologies are a way of living, of carving out a distinctive course through the world which confronts them. They are also ideologies which have arisen from a certain social milieu. Adherents may rarely recognise that they are engaged in acting out an ideology for this is left to the observer who is himself uninvolved in such a task. By a transmutation of secular events in the secular realm the adherent imbues such events with transcendent qualities.

NOTES TO CHAPTER SIX

[1] As far as I am aware no sociological work has appeared on the New Church. I have drawn on accounts of the movement from the following mainly historical reference.

L.B. DeBeaumont, *Spiritual Reconstruction and the Religious Unrest of the Age*, 1918.

J.S. Bogg, *A Glossary*: or *a Meaning of Scientific Terms and Phrases used by Swedenborg in his Theological Writings, Given in his Own Words*, 1915 Swedenborg Society.

Life of Swedenborg (revised by H.G. Drummond), London: New Church Press Ltd., 1956.

H.G. Drummond, *Degrees*, 1908.

Charles W. Ferguson, *The Confusion of Tongues*, London: Heinemann, 1929, pp.240-364.

The General Conference of the New Church Year Book, London: General Conference of the New Church, 1971-2.

John Goddard, *Wright and Wrong Unveilings of the Spiritual World*, New York, 1912.

Encyclopedia of Religion and Ethics, New York, 1922, edited by James Hastings, Article on "Swedenborgians".

Alan Grange, *The Structure of New Church Teaching*, Missionary Society of the New Church, London, 1963.

Helen Keller, *My Religion*, New York, 1927.

J.F. Molloy, *The Faiths of the People*, Vol. 2, London: Ward and Downey, 1892.

Dradoba Pandurung, *Reflections Respecting the Doctrines of the New Jerusalem Church*, 1879.

T. Parsons, *Outlines of the New Religion and Philosophy of Swedenborg*, 1876.

Cyriel O. Sigstedt, *The Swedenborg Epic: the Life and Works of Emanual Swedenborg*, New York: Twayne, 1952.

Julian K. Smith and William Winsch. *The Gist of Swedenborg*, Philadelphia: J.P. Lipincott, 1920.

[2] The family circle included the outstanding theologians of the day such as Bebel, professor of theology and church history at Strasburg, Sebastian Schmidius, translator of the Bible published at Strasburg in 1696, and Clodius, Hannekenius and Arcularius.

[3] His list of fourteen inventions included also a syphon, a sluice plan, a water-clock, a universal musical instrument and an air gun.

[4] His publications were many, and included The First Principles of Natural Things (1734), The Brain (1742), and Rational Psychology (1742). From time to time he also read "Memorials" to the Swedish Diet concerned with the finances of Sweden, a plea for establishing rolling mills, and so on.

[5] This is a gradual conversion, or what Berger calls alternation, by no means an unusual occurrence but one which exhibits much less drama than conversion of a sudden variety.

[6] Swedenborg only occasionally returned to secular activities, when for example in 1759 he read a Memorial for a return to a pure metallic currency. It was not long before the "Servant of the Lord Jesus Christ" had two attempts made on his life.

[7] I am indebted to the Rev. C.H. Presland, Secretary to the New Church Conference for supplying me with information relevant to this section. The source of this paragraph is a mimeographed sheet written by him. I alone am responsible for the interpretation.

[8] "And I, John, saw the Holy City, New Jerusalem, coming down from heaven." Rev. XXI, 2. The prerequisite for membership of the organisation of the New Church is public profession "of the Lord Jesus Christ as the only God of heaven and earth.... (and the members') intention with His help to live a life according to the Ten Commandments as His Laws". Swedenborg states that the quotation from Rev. "signifies the New Church to be established by the Lord at the end of the former church, which will be consociated with the New Heaven in Divine truths as to doctrine and as to life". Apocalypse Revealed, number 871.

[9] Mrs. M.L. Friend, manager of the New Church Press, has assisted me in piecing together these facts, although naturally the final form is my own.

[10] The New Church, operating from the same address as the former, in addition to the Church's adver-

tising, publishes a certain amount of material.

This bears out the pattern suggested by other sects, for example the Catholic Apostolic Church, the Exclusive Brethren, and the Witnesses. The Link servicing "isolated numbers" and "associates" of which there were 470 addresses in the Register in 1971, is also published by the New Church Press.

[11] Apart from the prgramme of conventional missionary activity the New Church has recently embarked upon a publicity campaign which consisted of five advertisements in the Sunday Times. For an outlay of Pounds 52 eight replies were received from Great Britain and one from Rome. Attempts have been made to sell copies of Swedenborg's writings of "Pop" Festivals and an approach was made to the "Free Communications Group", a commercial radio venture, which was discontinued when it was discovered that the aims of the radio groups were not in sympathy with those of the New Church.
The New Church is also engaged upon a complete translation of the Old Testament (now finished) and the New Testament. It also numbers among its organisations the New Church Sunday School Union, (1,099 children), the New Church Federation which is a youth movement with its own publication.

[12] Bryn Athyn and the Academy of the New Church, Bryn Athyn: Pa., 1903, p.27.

[13] A Brief Handbook of Information Concerning the Bryn Athyn Cathedral, Bryn Athyn: Pa., 1926, p.37.

[14] Alan Grange, Why the New Church is New, Scottish Association of the New Church (pamphlet).

[15] Swedenborg, Heaven and Hell, number 414.

[16] In many ways New Church doctrines, especially in an area such as this, laid the foundations for New Thought and Spiritualism (see below). Warren Felt Evans, one of the first advocates of the New Thought, was a New Church minister, and Thomas Lake Harris, the founder of the Brotherhood of the New Life, was a former Swedenborgian minister.

[17] I am relying heavily on Alan Grange, The Structure of the New Church Teaching, op.cit., pp.123-139, for what follows.

[18] Swedenborg, True Christian Religion, number 779-780.

[19] Swedenborg, Preface to Apocalypse Revealed.

[20] Swedenborg, Invitation to the New Church, posthumous publication section 43 and 52.

[21] Alan Grange, op.cit. en passent.

[22] Swedenborg, The New Jerusalem and its Heavenly Doctrines, section 252.

[23] Alan Granger, op.cit. p.131.

[24] Ibid. p.71.

[25] Swedenborg's "Intercourse with the Spiritual World" is alleged to be attested many times, and the best known are The Case of the Stockholm Fire (which Swedenborg "saw" with remarkable accuracy); The Receipt in the Bureau (when Swedenborg directed the searchers, again with remarkable accuracy, to a lost receipt belonging to the late M. de Marteville); The Dowager Queen of Sweden (when Swedenborg related a conversation which was known only to the Dowager and the dead Prince Royal of Prussia). New Church members are careful to point out that the truth of Swedenborg's writings is not dependent on these "miracles". The artist William Blake was a Swedenborgian for a time, and his work displays a visionary insight akin to Swedenborg's own.

[26] As late as 1909 Swedenborg's remains were transported by train to Dartmouth and then on the "Eylgia" to Carlscrona. His remains now reside in Upsala Cathedral.

[27] Servitus, burned by Protestants in 1553, pre-empted many of Swedenborg's ideas.

[28] Geoffrey K. Nelson, Spiritualism and Society, London: Routledge and Kegan Paul, 1969, pp.20. 50-54.

[29] See a similar argument in Robert H. Lauer, "Social Movements an Interactionist Analysis", in the Sociological Quarterly 13, Summer 1972, pp.315-328, esp. p.315.

[30] Egon Bittner, "Radicalism and the Organisation of Radical Movements", American Sociological Review, Vol. 28, No. 6.

[31] Ibid, p.930.

[32] Ibid, p.933.

[33] Peter Worsley, The Trumpet Shall Sound, McGibbon and Kee, 1968, has something to say on this theme. It is indeed true that the concept of Leadership

in social psychology has contributed a great deal to the idea that charisma might be best conceived of as an interpersonal phenomenon of role relationships within a group. Situational conditions are considered as increasingly important for determining the emergence and sustainance of a leader with charismatic "qualities".

[34] Op.cit. Lauer.

[35] Herbert Blumer, "Collective Behaviour" in Lee, Afred McClung (ed.), Principles of Sociology, New York: Barnes and Noble, pp.192-212.

[36] Lauer, op.cit., p.318.

[37] Lauer, p.320.

[38] Rudolphe Heberle, Social Movements: an Introduction to Political Sociology, New York: Appleton-Century-Crofts, 1951, pp.94-100.

[39] J.A. Geschwender, "Explorations in the Theory of Social Movements and Revolutions", Social Forces, 47, December 1968, pp.127-135.

[40] Werner Stark, The Sociology of Religion, Vol.II, Sectarian Religion, Routledge and Kegan Paul, London, 1966-1967.

[41] Luther P. Gerlach, and Virginia H. Hine, People, Power, Change: Movements of Social Transformation. Indianapolis and New York: Bobbs-Merrill, 1970, pp.79f.

[42] Bernard O. Brown, "An Empirical Study of Ideology in Formation" in Review of Religious Research, 1968, Vol. 9, part 2, pp.79-87.

[43] Ibid, p.80.

[44] J.L. Simmons, "On Maintaining Deviant Belief Systems: a Case Study", in Social Problems, Vol. 2., No.3, 1964, pp.250-256.

[45] Ibid, p.256.

[46] Peter Berger and Thomas Luckmann, The Social Construction of Reality, Garden City, New York, Doubleday and Co. 1966 edition, p.107.

[47] R.G. Collingwood, Metaphysics, Oxford: Clarendon Press, 1940, p.21ff. Within the terms of our own concern presuppositions underlie the external world. All statements, whether everyday or contained within an elaborate ideological structure, answer questions which themselves arise out of these presuppositions.

Such presuppositions are relative and historically influenced, and at times presuppositions are "con-supponible", that is, because they sometimes conflict it is also possible to hold them simultaneously. In this sense deviant or minority beliefs can be adhered to which appear at variance with beliefs and knowledge about the "real world" because each condition which gives rise to different presuppositions has its own distinction.

7 The organisation and structure of a secular value-oriented sect: the Socialist Party of Great Britain

Political sectarianism

It was probably Engels who first concerned himself with the study of religious sectarianism, if only to illuminate his concern with political movements and ideologies, and which he saw as embryonic political movements offering a parallel to the political sects of his day. Marx, however, in a letter to F. Bolte, dated 1871, expresses a slightly different emphasis when he writes: "...The International was founded in order to replace the socialist or semi-socialist sects by a real organisation of the working class for the struggle...(it)...could not have maintained itself if the course of history had not already smashed the sectarian system. The development of the system of socialist sects and that of the real workers' movements are always inversely proportional to each other". He sees the historical justification of the existence of sects as an impediment to the existence of a true working class movement. The reification of such a working class movement means that when it reaches maturity all sects become essentially reactionary. But to both men religious sects were seen as organisations blindly striving for political consciousness although contained within the blinkers of religious commitment.

Furthermore, both men tried to establish the relevance of primitive Christianity to mass movements, which might be found in the secular and early historical Christian differentiations. From an account of the social revolutions in medieval Europe, such as the Chiliast peasant risings, the Taborities of Bohemia, and Muenzer's Saints, they pinpoint a sense of urgency in social conditions and the emergence of a revolutionary disposition.

Trotsky went a step further in refining the usage of the term "sectarian", as someone who is engaged in a discord with reality and who -

> sees an enemy in everyone who attempts to explain to him that an active participation in the worker's movement demands a constant study of objective conditions, and not haughty bulldozing from the sectarian rostrum. For analysis of reality the sectarian substitutes intrigue, gossip and hysteria.[1]

The collective behaviourist tradition extended the analysis of sectarianism as applied to religious and political movements. Thus Scipio Sighele wrote that "Any faith, whether it be Islamism, Buddhism, Christianity, patriotism, anarchy, cannot but pass through this sectarian phase".[2] Kurt and Gladys Land[3] adopt a conflict theory of sectarian development, seeing the sect as having other than a merely religious connotation. In a predominantly religious milieu religious sectarianism will arise: in a political milieu there will be a surfeit of political sects. Basically, they view sectarianism as consciously separated, "a compensatory brotherhood of the select, dedicated to an unconventional authority of their own choosing". Psychologically, sectarian adherents are viewed as misfits or outcasts who affiliate to the conflict associations in the hope that they will be rewarded by a more genuine experience or ultimate cause.

Coser[4] suggests that sectarianism, whether manifested in political or religious expression, will exhibit a number of similar patterns of behaviour: "I am quite aware that there are fundamental differences between say, a fundamentalist sect and the political sects of the modern socialist movement, but the struggle of sects as such breeds certain common patterns".

Certainly, there seems to be a respectable precedent in the application of the term "sect" to cover movements and groups that are to be found outside not only the Christian tradition but also to those which are not religious at all. As O'Dea has said, "Groups with situationally transcendent ideals seem to display similar organisational types".

For some the antecedents of both socialism and Marxism are to be placed firmly in the camp of the religious, not only because both share a similar dignity and prestige "normally reserved for a religion" but also because there may be the possibility that, with a convert, there has been a

> transfer of religious fervour into a social conviction. Although there has been an exchange of content, it is possible that some characteristics will remain - <u>either</u> the modes of depersonalisation: enthusiasm in conformity, the effervesence of a certain eschatology, exclusivism, certainties by proxy, intransigence, even some-

> thing like a historical Manicheism. Or, on the other hand, modes of personalisation might survive: appetite for coherence, universalism, longing for the ideal of abnegation, devotion to a conviction, development of the scale of participation, etc.[5]

Desroche goes further and compares the German Reformation with the Proletarian Revolution:

The German Reformation	The Proletarian Revolution
1. Transformation of the layman into priest.	1. Transformation-into-man of this layman-turned-priest.
2. Political emancipation of the ruling classes.	2. Political emancipation of the masses.
3. Expropriation of the Church's wealth (by secularisation).	3. Expropriation of the whole of capitalist property.[6]

And in a letter to A. Ruge Marx writes:

> The reform of consciousness consists solely in making the world aware of its own consciousness, awakening it from its own inward-looking phantasies and explaining its own actions to it...our aim can only be to invest religious and political questions with consciously human form...A confession (Beichte) is required, nothing more. To have its sins forgiven, humanity has only to declare what they are.[7]

There can be little doubt, then, that not only was the young Marx imbued with a religious approach to society which continued into later life, but that there is a whole wealth of information which connects the beginning of Marxism and socialism with religious and sectarian movements. For the purpose of this section we will define a political sect as a small political group in which membership is voluntary and conditional upon some mark of merit - perhaps understanding the group's teaching - upon the basis of which association can arise. It is small and exclusive, and in conflict with the larger society.

Contemporary attempts at an analysis of political sects and their ideologies

Further substantiation of the claim that secular groups have often displayed an organisation and structure either very similar to transcendental groups or in many ways inseparable from them may be seen in our references to Black Nationalism and Communism, which have already been suggested in this light. For example,

In Holland and Sweden, recent studies have shown that the Communists are strongest in regions which were once centres of fundamentalist religious revivalism. In Finland, Communism and revivalist Christianity often are strong in the same areas. In the poor Eastern parts of Finland the Communists have been very careful not to offend people's religious feelings. It is reported that many Communist meetings actually begin with religious hymns.[8]

A more systematic and sustained study by Murvar[9] suggests that religious and revolutionary groups with messianic characteristics in Russia generally share the same basic doctrines, beliefs, attitudes and behavioural expectations. Furthermore, they exhibit an "extremely close structural and functional interdependence", particularly in the three following messianic doctrines which he singles out for detailed analysis:

1. Millennialism (chiliasm): an immediate millennium.

2. Twain cosmogony, and a corresponding division of humanity into "children of light and darkness".

3. Spiritual, political, and economic monism and/or collectivism; obviating the distinction between secular and spiritual, etc.[10]

Furthermore, revolutionary and religious Messianism also share a number of other attributes, namely, an eschatological search for a total truth; exclusiveness; absolute intolerance; totalism; sainthood and asceticism; extremism; martyrdom and martyrology; culpability and expiation before "the people"; isolation and estrangement from reality; and apocalyptic prediction.[11] An analysis of the Dekabristi, Zapadniki SPV (Slavophili, Pocveniki, and Vostocniki), and Narodniki doctrines with religious messianism reveal many similarities, and such similarities were conducive to social change both prior to 1917 and afterwards. "Indeed Marxism itself - an essentially Western doctrine - was successfully Russified by these messianic influences."

Others, also, have drawn parallels between secular and transcendental value-orientated movements.[12] For example, Schwartz[13] talks about persons "motivated by political ideologies often use a vocabulary which has ritual overtones. They talk about 'cleansing' the social fabric of 'impure', 'subversive', or 'reactionary' elements. And they propose to 'save' the misguided or misinformed from alien influences and, in the process, to regenerate dormant virtues in the service of a noble cause. For these ideologists, the exhortation to attack or protest the prevailing social order is validated by external and self-explantory values - that is, by spiritual values. In this manner, a religious rationale is embedded in ostensibly secular ideologies".

One recent and interesting example is the growth of a group of professionals whose function it is to "care" for others in society. Sometimes known as the "secular priests" these professionals such as psychotherapists, hold certain assumptions in common about what makes man "tick" or society "work". In terms of the work that they do they "administer help to others through the medium of intimate personal relationships". Moreover, they entertain particular "metaphysical beliefs and moral affirmations" which amount to "a creed, a faith, an ideology".[14]

In a society which is increasingly technocratic it is perhaps no small wonder that agencies have arisen and are arising which are ostensibly secular in origin and characteristic yet are imbued with a quality of sacralism in the very practice of what they do. Thus the emergence of psychotherapeutic ideology in what some have called the "professional service society" and others the "psychological society" arises to fill what yet others have termed the "existential vacuum" or the "crisis of identity". Perhaps this is misplaced functional naivity, but it nevertheless remains the case that cognitive and affective "wants" are increasingly being catered for by other than sacral agencies.

Certain key concepts such as "salvation" can, also, be capable of both secular and transcendental significances. It is quite common nowadays for scholars to talk of secular transcendence. Kee, for example, talks of faith, hope and love, the cross and resurrection, all in terms of transcendence.[15] Encounter and therapy groups are specific examples of secular value-orientated groups which, from time to time, can become transposed with transcendental qualities.[16]

The ideologies of both secular and transcendental movements appear to share many characteristics in common. Essentially they are dispositional states or propensities to action located within the structure of society rather than possessing any unitary or intrinsic characteristics. On the whole, persons locating generally similar positions in the social stratification process will be more liable to adopt similar ideologies. The ideologies themselves are exaggerated models of the "real" world; bounded stocks of meanings which any individual can draw upon in his interpretation of the world. In one sense they are deviant ideas; distortions of the world which are aided by an emotional investment of the adherent or believer in the ideology. It is often quite possible to make the same circumstances <u>in</u> the world capable of a judgement which contains no hint of this distortion.[17] The Marxist would regard ideologies as representations of false consciousness which distort the scientific perspectives; "men and their circumstances appear upside down as in a <u>camera abscura...</u>".[18] In other words, an ideology appears as such only to those who are standing outside any emotional involvement in it, and who are in essence taking a reflective look or analysis at the set of beliefs. Moreover, the set of beliefs which constitute an ideology always present a <u>total</u> picture of the world. Schwartz[19] argues that adherents to an ideology commit themselves to it not on the grounds of belief in separate propositions but in relation to

> three analytically distinct dimensions of an ideology. First, ideologies are recognisable as such only to an observer. Second, ideologies not only arise in a particular social milieu but also often have a decisive impact upon it..... ideologies dramatise the universal dilemma between the pressures of environmental constraints and the possibilities of cultural innovation in social life. As a collective attempt to resolve common problems, ideologies are free, within certain limits, to impose their own interpretations on the sources of man's ills. Primarily, ideologies locate and provide remedies for human suffering, and perform a function of reducing uncertainty. They are essentially mobilising agencies.[20]

Third, ideologies become believable under particular circumstances only when the potential believer is attracted to the ideology, and when the circumstances of his social life are such as to orientate him, with the appropriate disposition, in the direction of the ideology. Ideologies are shared, articulated responses to situations within the external world which give a "total" explanation. The embracing of an ideology can carry with it into the real world particular constructions of meaning, whether this ideology is religious or secular, which somehow shape that real world for the adherent in a way that is different from someone who does not embrace that ideology.

Throughout several works that attempt an explanation of ideology, for example Mannheim, Durkheim and Marx, we come across the idea of ideology as a stock of meanings which assist individuals to interpret their world, that is ideology is conceived of as a stock of meanings to which particular individuals have specific access. Such ideas as access to, and palatability of, ideologies are determined in a broad sense by the social position which we occupy which in turn sets the picture of reality which we entertain. Consequently, broadly speaking groups of people occupying the same position in the social structure may be seen to hold roughly similar views about events and phenomena in the world. Different social groups in society hold differing ideologies and while this on a superficial viewpoint can make for an interesting and diversified society it can also present real difficulties in terms of agreeing with or even understanding ideologies which differ from one's own. These difficulties can become more acute depending on whether the ideology is a closed or an open one. Ginsberg designates the former as "self-contained systems demanding all or no commitment, of the kind demanded by, for example, Bolshevist communism". The latter are ideologies because "they are often criticised from within and...they also learn from each other".[21]

In societies which are becoming increasingly secularised and pluralistic individuals and groups find legitimising forms in new and often strange areas. Periods of social stress and unrest are particularly fertile times for the emergence of new movements. For example, the early 1830s saw the growth of several new variations on a basically Christian ideology, apart from the Catholic Apostolics. Schwartz sees "convincing parallels between the development of chialistic and apocalpytic religious visions in the earlier phases of Western history and contemporary revolutionary doctrines of both the right and the left".[22] These new movements

> mobilise their followers through appeals to transcendent moral values and thereby infuse their political programmes with a sacred aura...Like their religious counterparts, these ideologies interpret ordinary human affairs in terms of categories of absolute good and evil...Whether they are politically right or left, these ideologies seek to generate a moral crisis.[23]

In secular and pluralistic societies the notion of salvation undergoes a transformation.

> Here the idea of personal salvation is understood as a persistent effort to preserve the self against the vagaries of time and nature and, at the same time, to intensify its qualities which partake of the eternal and the divine...salvation involves attachment to collectivities (often pseudocommunities) which translate the individual's fear of existential meaninglessness and annihilation into a search for enduring sources of value and for modes of transcendence into these spheres of being.[24]

Schwartz quotes the work of Rieff[25] whose study of the schismatic followers of Freud has demonstrated the emergence of a new faith where the individual has freedom to "explore all the possibilities of his experience, save those which would damage the self or impair the autonomy of the ego". Various encounter groups, therapy groups, "mind-cure" movements such as Dr. Low's and Norman

Vincent Peale, some of which we refer to elsewhere, are all variants of this new expression of legitimation to human existence.

> All troubles, all anxieties, all depressions dissolved. A euphoric state of unity and one ness between all the participants transcended such considerations as status, wealth or sex. Everyone felt good about everyone else.[26] Here, then, salvation involves a process in which the seeker acts as a sentient siphon for the multitude of cosmic forces and prosaic events which affect his life. Salvation, then, arises out of a continuous openness to experience and is renewed by freely sharing the fruits of this enriched experience with like-minded others.

Shepherd[27] argues much the same way for a validation of such pop festivals as Woodstock in 1969, drug taking among the young, and so on, as genuine religious experience not confined within the old boundaries of theistic commitment.

One of the consequences of modern industrialised societies in the Western world (as well as emerging nations and states) is the production of belief systems which are highly privatised and which some might designate as unbelief.

It may well be that the conventional (and primarily Christian) preoccupation with the categories of "belief" and "believers" is no longer applicable to demarcate this category from "non-believers". Consequently the term "religious" can be applied to all that offers man room for some sort of maneouvre within the confines of a society constantly bounded by impersonal and technological agencies, the triviality of urban living, and an almost inevitable prospect of modern man living his life in a purely rational situation with a rationalised framework of action. Some sociologists (notably Parsons and Bellah) have argued cogently for movements which manifest themselves in secular love. The campus movements in America, for example the Peace Corps, have offered alternatives to traditional religion for the expression of a personal and extremely social vision. Similarly, the campus youth in their search for identity and spirituality in an increasingly materialistic society have embraced systems of commitment and ultimacy which are the equal of institutionalised "orthodox" religions. How else can we view the hippie cult, and the Rama Krishnam movement for example? But there is nothing particularly new here because movements embodying equally valuable social insights quite explicitly rejected transcendent belief systems throughout the nineteenth century and even earlier, e.g., Comte's "religion of humanity". These movements[29] and in particular the new "religion of revolt" in America and Britain, with their deep and sincere commitment to an ultimacy of values expressed in an abhorance of war and the oppression of poverty, are explicit, equally, in their vehement rejection of a value-system based on a theistic interpretation of the world in favour of a radical orientation to the conditions inherent in this world, the ultimate purpose of which is the belief in the feasibility of the creation of a community of love as an integral part of the everyday world.

Ideologies, both transcendental and secular, are singled out by their greater explicitness, greater internal integration or systemisation, greater comprehensiveness, greater urgency, and higher intensity of concentration.[30] Furthermore, both types of ideology are concerned with the sacred, seeking to "sanctify existence by bringing every part of it under the dominion of ultimately right principles". Shils also argues that the growth of secularisation has been accompanied also by the growth of literacy and general education which has had the result of widening receptiveness to ideological beliefs. Sharers of ideologies may come to form an ideological primary group which tends to erect boundaries of demarcation, procedures for the management of conduct and belief, etc., these primary groups becoming important bearers of ideology in society. Ideologies, Shils stresses, "intend either the disruption of the central institutional and value systems by conflict with them or the denial of the claim of these systems by withdrawal from them in the former case. Ideologies aim at 'total' replacement".

The Socialist Party of Great Britain, or SPGB, is a movement which firmly expresses a secular value-orientated ideology. Unlike the overtly sectarian and transcendental groups, or the secular value-orientated therapeutic groups, the SPGB is a political group expressing an articulate irreligious ideology. But despite this, the SPGB is a movement which displays distinct similarities to religious sectarianisms both in its organisation and belief, and also to other value-orientated and explicitly non-transcendental groups such as Alcoholics Anonymous.

The Socialist Party of Great Britain: its origins and growth

The fragmentation of political groups, both of the Left and Right, has been equal, almost, to that displayed by the breakaway of religious groups from the mainstream tradition. Thayer, for example, in a popular account lists nearly one hundred.[32] Reid, in her book Ultra Leftism in Britain,[33] gives an account of dozens of left wing sects. The SPGB emerged from the political turmoil of the Left during the latter part of the nineteenth century. Towards the end of that century, when society developed a superficially increased tolerance to socialist movements the Social Democratic Federation was the most advanced of the English radical parties of the Left. Dissatisfaction with the SDF's reformist policy had already led to a breakaway of a group which formed the Socialist League, comprising such members as William Morris and Eleanor Marx. Blessed with the sanction of Engels the Socialist League eventually came under the control of the anarchists, and abandoned parliamentary action. During the early part of the twentieth century the SDF was again subjected to fierce criticism from within on the grounds of the movement's policy of reformism, leadership, private agreements and political trading. "Their (SDF's) efforts, however, were paralysed by the power, influence and secret arrangements of the official leaders".[34] This revold led to the formation of the Socialist Labour party in 1903 and the SPGB in 1904. All the revolts had been aimed primarily against the man instrumental in setting up the SDF, H.M. Hyndman, who "seemed to regard it as his personal possession and reacted to any criticism in a haughty and aristocratic manner".[35] Hyndman was representative of the older, middle-class group of men who controlled the SDF and who advocated a policy of including all in the party who protested against social injustice.[36] This was felt by younger members to be insufficient, and they advocated maintaining a clear perspective in relation to the social class basis. Yet another point of protest arose over the official party organ, Justice, which was edited by H. Quelch although owned by a private group over which the members had no control whatsoever. The SLP was based on an American organisation of the same name. Its origins followed the appearance of a paper in Scotland in August, 1902, entitled The Socialist, published by the Scottish members of the SDF. The third issue was an attack on the leaders and policies of the party, and was signed "Impossibilist", following Quelch's nickname "The Impossibilists", which he had bestowed on the militant younger members. It was this group of Scottish Impossibilists who formed the SLP. From the beginning, at least according to the SPGB archives, the SLP became lost in industrial unionism; "in fact it soon became apparent that the members of this party had only really changed their ideas; Hyndman, Quelch and company were deposed, and De Leon and Connolly took their places".[37] The situation escalated, and a number of letters critically directed against the party and its leaders were printed in Justice, together with appropriate replies by Hyndman, Quelch, Lee and Max Boer. A banquet was given by Hyndman, to which all political persuasions were invited, which was criticised by members.

The third revolt against the Hyndman group and the SDF, the one which led to the founding of the SPGB, was organised by younger members considerably better informed about Marxist political and economic theory. Conferences were showing a very high attendance, branches were closed down and many members expelled by the Executive Committee of the SDF. "Throughout 1903 the volume of criticism against the autocratic attitude of the Executive Committee and its compromising policy grew stronger and more articulate". At the beginning of April, 1904, the Conference brought matters to a head.[38] By a vote of 56 to 6 the Executive Committee voted that the Impossibilists be called upon to apologise for their behaviour to the Conference and subsequently refrain from future conduct of that nature. No apology was forthcoming and two of the members, J. Fitzgerald and H.J. Hawkins, were expelled. "The official group complained that the Impossibilist movement was a campaign of calumny and intrigue against old and experienced members and, therefore, against the entire body."[39] When Fitzgerald's Watford branch demanded an explanation, they also were threatened with expulsion.

A special meeting, mostly of the London delegates, signed the following statement:

> The adoption of any uncompromising attitude which admits of no arrangements with any section of the capitalist party; nor permits any compromise with any individual or party not recognising the class war as a basic principle, and not prepared to work for the overthrow of the present capitalist system. Opposition to all who are not openly and avowedly working for the realisation of Social Democracy. A remodelled organisation, wherein the Executive shall be mainly an administrative body, the policy and tactics to be determined and controlled by the entire organisation. The Party Organ to be owned, controlled and run by the Party. The individual member to have the right to claim protection of the whole organisation against tyrannical decisions.[40]

The SPGB was founded on June 12th, 1904, and the organisation and structure was firmly designed to make sure that what had happened in the SDF would be impossible within the new party. Conference policies would be executed by the Executive Committee, elected by the whole of the membership, which was required to report to the membership on two occassions a year. The Socialist Standard for 1954, the fiftieth anniversary issue, had this to say:

> Owing to the bitter experience of the undemocratic methods of the Social Democratic Federation the new party framed rules that gave the whole of the membership complete control of the organisation, and, in order that workers could be under no delusion about the aims and activities of the Party, all meetings, whether Branch Meetings,

> Executive Meetings, or Conferences, were open to the public; anyone was free to enter these meetings and listen to the discussions.
>
> This was a revolutionary departure from custom and a severe blow to the cult of leadership...

The first issue of The Socialist Standard appeared in September, 1904, and has appeared monthly ever since. Adhering steadfastly to the Declaration of Principles and the Object which appeared in that first issue, the SPGB has excluded from membership all those who hold allegiance to any other political party or to any supernatural belief. It has insisted on a majority capture of political power, opposition to reform policies, war, leadership, singlemindedness of members, and an advocacy of public scrutiny and democratic organisation. The SPGB has also maintained a pristine purity in not recognising any other "socialist" party.

Organisational structure

Membership of the SPBG has never been very great, and present membership figures are not high. In 1904 its membership was approximately 116; in 1914 approximately 464; in 1941 approximately 700; in 1948 approximately 1100; in 1955 approximately 700. Three "crisis points" occurred; in the years during the First World War when members were opposed to the conflict; the mid-fifties, when the advent of television and a newly elected Tory Party meant a veritable end to outdoor meetings and a certain amount of apathy was introduced; and the departure in 1960 of two members prominent in the SPGB who resigned together with thirteen members from the same Branch.[41]

Rolling's study of the SPGB highlights six facets of consistency which characterise the group and to a large extent had determined the organisational structure.[42]

1. Consistency of attitude - the group has never altered its basic view as enshrined in the Object and Declaration of Principles, and furthermore its literary output has changed very little. "In 1955 the 'D of P' came under severe attack from the 'heretics', but it survived intact; more, continued agreement with it became a criterion for continued membership."

2. Consistency of membership - the formal criteria is an interview or written test which establishes the applicant's knowledge of the SPGB's case.

3. Consistency of propaganda - all articles appearing in the Socialist Standard are vetted, as are representative speakers.

4. Consistent rejection - of all "capitalist" features of modern societies, with a concomitant emphasis on

a total embracing of socialism.

5. Consistent hostility - to all other parties, socialist or right wing.

6. Consistent democracy - every record is available for public scrutiny. The SPGB represents a total ideology based on Marxist principles which encompasses all features of existing society, and it "plainly represents an extreme form of an attitude found in more confused and less consistent forms much more widely: that of opposition to the established order of society".[43]

The membership composition appears to consist of skilled manual workers, with a significant number of building trade workers "and clerks with an elementary education, subsequently self-educated, with a sprinkling of businessmen, managers, senior civil servants...and a somewhat larger number of teachers and self-employed people". According to Rolling, the "more conventional process of entry into the professions through secondary education followed by university or professional training is, with few exceptions, undergone only by the second generation, the 'cradle socialists' who have benefitted from the self-advancement of their SPGB fathers".[44]

The Executive meets every Tuesday at the Head Office in Clapham High Street, and apart from that Branch meetings occurred on a regular basis in seventy areas in 1972: (Birmingham; Bolton; Camden; Ealing; Edinburgh; Glasgow; Hackney; Haringey; Lewisham; Merseyside; Manchester; Mid-Harts; Redbridge; South West London; Swansea; West London; Westminster). This compares with thirty-one Branch meetings in 1954: (Birmingham; Bloomsbury; Bradford; Brighton; Camberwell; Croydon; Dartford; Ealing; Eccles; Fulham; Glasgow (2); Hackney; Hampstead; High Wycombe; Islington; Kingston-on-Thames; Lewisham; Leyton; Manchester; Nottingham; Paddington; Palmers Green; St. Pancras; N.W. London; Southend; Swansea; Tottenham; West Ham; Wickford; Woolwich), and twenty-one in 1931: (Battersea; Becontree; Birmingham; Central; East London; Glasgow; Hackney; Hull; Islington; Leyton; Manchester; Paddington; Sheffield; Southend-on-Sea; Southwark; Tooting; Tottenham; Walthamstow; Watford; West Ham; Wood Green). Apart from the Branches there also exist SPGB Groups and Information Centres, and lectures and special meetings throughout the country embracing such topics as "Why a Million Unemployed", "Chronic Crisis of Capitalism", "Primitive Communism", "Iron Law of Oligarchy - Is Michels Correct".

Companion Parties, that is those who have adopted the Object and Declaration of Principles exist in Australia, Austria (Bund Demokratischer Socialisten), Canada, Ireland (W.S.P.I.), New Zealand and U.S.A. Some isolated groupings occur in other parts of the world, mostly sustained by ex-SPGB members (e.g., in Scandanavia, India, Malaya and South Africa).

As an organisation the SPGB has a certain rigidity of control stemming from a centralised authority in the form of the Executive Committee. The authority displays itself in the vetting of the Socialist Standard material and representative speakers.

> The material hasn't changed since we began. If you have the correct view how can the Socialist Standard change? It's only people who change.

It also maintains a watching brief on the observance of the Object and Declaration of Principles. Consequently, because of the Party's early history and origins there are built-in mechanisms for safeguarding any excessive autocracy on the part of the Executive Committee. Branches have their secretaries and elected representatives on the Executive Committee and the Conference. As in many of the other value-orientated groups organisational characteristics of a normative kind are exhibited by an intense level of commitment, and by an internalisation of directive stemming from the Object and Declaration of Principles which are regarded as the sole legitimate directives or definitions of socialism. Like Alcoholics Anonymous, there is no charismatic leadership of any kind, although the inevitable process by which long standing members are regarded with affection and perhaps some pioneer aura is present.

Despite a completely open and rational type of organisation the SPGB does not in fact keep very accurate records. The ones we had access to were stowed randomly on the shelves of Head Office or sitting in piles on the floor. No one asked knew the exact number of members, which we estimated to be approximately six hundred but some have suggested seven hundred and eleven. There are some two thousand four hundred living ex-members. The Head Office is situated in a rather run-down house at the back of Clapham High Street in South West London, and other SPGB branches are located in most of the major cities in Britain, although membership in each of these branches will not amount to thirty at the most. Certainly the Head Office is the busiest and most active centre of the whole movement. It seems that on a rough estimate about a quarter of the membership is recruited through a family connection, more by marriage than by family descent. Others come to join through replying to advertisements in the Tribune and other papers, or simply through the agencies of friends.

> I used to be a Catholic. I wrote off for the literature and my eyes were opened. I realised it was the only way. My wife was a Catholic,

> too, but eventually she also came to believe in the truth of the SPGB.

We gathered that many of the "old guard" thought it far easier to become a member of the Party today than it was in the early days. Accounts seem to suggest that individuals once had to fight to get into the Party. Perhaps it is worth observing that generally those first in a group will invariably try to suggest that younger and later members have it "easier" in many ways.

Attrition from the movement appears to be due to a number of reasons. Firstly, there appears to be sheer apathy resulting in a lapse of subscription and subsequently membership. Secondly, it appeared that the demands of marriage and many of the responsibilities connected with child-rearing and full-time occupation tended to draw members away from the movement. There are other reasons, too:

> I'm not really committed as regards any activity now although a few years ago I put quite a lot into the movement. My time now is so full. I made the time then, and suffered during making it, you know. I'll be honest. We lost a member recently. He was making the sole effort on his own. He was a very good member, too, and used to be there on a Saturday on his own. He naturally became disillusioned.

We had difficulty in gauging the social composition in terms of the educational and occupational distribution of membership without the means of a questionnaire. Members certainly do not appear to be aggressive or misfits or in any way eccentric, but stem from the upper working and working class. Skilled workers, and in particular building workers, predominated. Unlike other political sects there is little evidence of any young intellectual element drawn from the universities, and they are more likely to be primary school teachers with a non-university background.

The difficulty of reaching any accuracy with regard to the career patterns and educational backgrounds of members is enormous without access to detailed and objective information. Members tend, it appears, to be drawn from skilled and manual workers and those with a minimum of education, although the trend, one would expect, might, with the rise in national educational standards, gradually be towards a decline in this concomitant.

An incident which emphasises the role of the Socialist Standard in the application of principles and theory is recalled as follows. An argument was in progress which concerned the involvement of the SPGB in support of the Civil Rights movement of the American working class negro. Basically, support for the movement was not forthcoming on the grounds that the character of the Civil Righters was essentially capitalist, and that the Civil Rights movement failed to understand the present state of American society and the social forces tending towards the abolition of the movement. One of the protagonists remarked as follows:

> The next time I heard from him he conceded that he had overlooked the real points of the issue, and that the Socialist Standard had set him straight.

As we have mentioned, there is no charismatic leadership of any kind for the movement is very wary of any inclination towards a "cult of personality", from which it suffered in its early days. Indeed, the group appears to bend over backwards to avoid any highlighting of its members as individuals. They reported:

> Dead comrades are treated less inflexibly; an obituary note in the Socialist Standard, perhaps on special occasions a rememberance of those who were outstanding by their labours for Socialism. The rest is self-effacement.

Recruitment techniques are virtually non-existent:

> We only recruit convinced socialists. We don't have a group in mind and that's why our numbers are so small. If they are socialists we will have them.

Some twenty years ago it was quite common for a number of representative speakers to take their places on street corners or in Hyde Park and address the passers, as well as holding public debates. Now the movement seems reconciled to a policy of non-recruitment although it is fastidious in answering postal inquiries and distributing literature by post. Currently, the programme of representative speakers covers only Hyde Park, Lincolns Inn Fields, the outside of Charing Cross Tube Station, Bromley, Earls Court, Edinburgh and Glasgow. One of the groups approached had been disrupted by members of the National Front and forced to abandon the premises they had used for over three years.

The application of the Object and Declaration of Principles and the theory in which they are incorporated, apart from the watchdog committee of the Socialist Standard, is apparent when an individual applies for membership and is required to answer questions on the party policy:

> We can tell that they are socialists by asking them selected questions.

> There are no other socialists. We are the only socialists. There are no other socialist groups. Any other alleged socialists are merely advocating reform. They are not advocating a revolution in society.
> If they are in any other party they cannot be socialists.

All members are encouraged to write for the Standard but no views are in fact contained in the publication which does not accord with party policy:

> The Object and Declaration of Principles relate directly to society, from start to finish. They haven't changed because the facts of the situation haven't changed, have they? It's the same society as when the Object and Declaration of Principles were framed. It operates in the same way.
>
> It's still the capitalist society. It's still capitalism but there's more state capitalism and private ownership. Like the basis of capitalism is capital labour and commodity production. They haven't changed at all.
>
> Only the superstructure of society has changed. The basis is the same as it always was. Some people say that the language is archaic and old fashioned but I don't see it is.

The SPGB is bound to view that it is the supreme and true example of socialism, and it is one of the few ultra-left groups, with the exception of International Socialists, to relegate the triumphs of the October Revolution to the area of "State Capitalism" as manifested in the Soviet Union. All other forms of socialism, other than those enshrined in the Object and Declaration and political parties, are unrepresentative of true socialism:

> The SPGB are the only people propagating the ideas of socialism. We are not the guardians, not quite. There may be hundreds of thousands of people who are socialists and don't know it. The Labour Party, International Socialists, the Labour League, name any one. They are just misusing words. The Communist Party misuses the main word in their name. Communism and socialism; the two words are synonymous to us. When it (socialism) comes it can only exist on a world-wide basis.

The SPGB is completely committed to remaining a totally distinct and separate organisation even if this means its final dissolution:

> Capitalism causes most of the evil in the world.

> We can't do other than propagate our ideas and by propagating them, when we finally succeed, we are out of existence and no longer have a function. The party itself is unimportant once socialism comes. It's only an organ of propaganda. When a majority understands capitalism, understands socialism, and desires change, the change will come. Until a politically conscious majority desires socialism it will never come. The SPGB will phase itself out of existence, which we are all quite happy about. We want socialism, not an everlasting SPGB.

Convinced that it holds the key to a scientific and rational truth in relation to Marxist dialectical thought and is in a unique position to unmask the illusions of the bourgeois world, it steadfastly puts forward a consolidated moral platform from which to overthrow the forces of "capitalist evil".

> We object to them calling themselves socialists just as we object to a communist calling himself communist. Half our time is spent convincing people what words in the English Language mean.

> We have the only interpretation of socialism. Yes, that is what we are saying. Anyone who advocated anything else other than socialism as we talk about it, our brand, is not talking about socialism. Do <u>you</u> think there's any other kind?

It further purports to put forward a unique perspective on the reality of the world, believing that under the mask of capitalism others are seeing through a "glass darkly":

> We are completely committed to socialism. In terms of commitment by <u>effort</u> a lot of us haven't the time. It's a conflict between living your life and propagating socialism. I'd be delighted to win the pools, then I'd be able to devote my whole life to this. I'd have the biggest van I could find and put placards on it and parade up and down. Because it could become fruitful because you could devote your whole life to it.

In terms of a moral elitism the SPGB sees itself as the vanguard of the true socialism and whilst spurning the bourgeois moral system upholds the "greater" morality by being more attuned to the socialist moral system. Application of socialist principles becomes increasingly difficult in a world which embraces capitalism at the expense of socialism:

> You can't live socialism. How the hell can you

live as a socialist when you are living in a capitalist society?

The nature of the member's involvement in the sect is such that it offers them many opportunities for social and intellectual fulfilment without ever having to venture outside the programme which the party offers in the form of weekly meetings:

> I believe in it. I think it's right. But I don't think it will ever happen, either.
>
> It's the ultimate in progress, you know. Society is progressing all the time. Socialism is the ultimate of everything and it will eventually come. It will come quicker through the SPGB.
>
> I am completely satisfied with the SPGB. In terms of my work it makes you realise I am being squeezed like an orange.

These weekly meetings sustain interest and various activities, recreational and intellectual, are extremely satisfying to the already committed. There is a totality of commitment, and an almost fundamentalist creedal devotion:

> According to the Declaration of Principles we must be hostile to all other political parties. The only people we are not hostile to are the uncommitted, by far the vast majority. The committed people in this society who are members of political parties are very small indeed. Even including the Labour and Conservatives they are still a small minority. Although Labour Party members may agree with fifty per cent they don't agree with a hundred per cent. And that's the whole point.

The Socialist Party of Great Britain and religion

The SPGB has been in consistent opposition to religion and the supernatural since its inception. In keeping with a fundamentalist historical determinism, the movement maintains that religion is a social product. The supernatural or transcendent conception of God is the result of man projecting essentially anthropomorphic imagery. Reactionary and divisive influences in society will fall beneath an increasingly scientific conception of society and the world. From time to time the Socialist Standard expresses the view that no one who entertains religious views is eligible to join the World Socialist Party and that "a criticism of religion must be tied to a criticism of society: only Socialism in fact would end the need for religion".[45]

Despite a sustained opposition to institutionalised

religion and belief systems, the SPGB evidences concern over moral and social issues. In fact it was a distinct feeling by some members that the Party was not sensitive enough to human suffering and misery which led to the resignations in 1960. Morality is the expression of quite definite and historically constituted economic and classic interests.[46] Again, the Party and its Companions frequently quote crime statistics[47] as indicating further evidence that such is the result of capitalist society.

A sectarian aspect of the Socialist Party of Great Britain

One serious attack on SPGB orthodoxy appeared in the series Towards One World, No. 3, Aspects of Religious Socialism, by Peter Lawrence,[48] which although not an official publication represents an intelligent criticism of SPGB orthodoxy as depicted by some of the more conservative members of the Executive Committee. Lawrence singles out a "solid phalanx of Party orthodoxy which seems in an irrational way to resist any kind of change". The result of this over some twenty years "is that the Party is less political in that it says less, is less influential, and is taken less seriously". By an intelligent comparison with an overtly sectarian group, Jehovah's Witnesses, Lawrence identifies "a strong element of religious socialism in the Party's make-up". He suggests that both the Witnesses and religious socialism are sectarian examples of alienated man sharing four similar characteristics:

1. dissociation from society;

2. sudden transformation of life;

3. retribution;

4. belief in perfect society.

A proposal at the 1971 Conference suggested adding to the Object and Declaration of Principles a note to the effect that they were written in 1904. This was resisted by the orthodox because "to the religious socialist, attempts to modify Party practices in line with better communication must be vigorously opposed, since they are seen as a threat to the group basis on which he at present attacks society".

Religious socialism shares with many sectarian groups of a transcendental nature "withdrawal from society, hostility, cataclysmic change, retribution and perfect world...The function of therapy is also clear. Such organisation provides a bulwark against authority, common association and the consequent relief of misery and stress". Despite a necessary superficiality (this is contained in a pamphlet) Lawrence's document is interesting on two counts. It is first of all a reaction, an "heretical" reaction, to a narrow and unyielding fundamentalism manifesting itself in a fatalistic acceptance of events in the world. The Catholic Apostolic Church in many ways displayed much the same reluctance to engage with the world, and events were taken as but further examples of the ideology's justification.[49] Secondly, the writer draws attention to the possibility of parallels existing between secular and transcendental value-orientated groups.

Party ideology as represented by the principles as a parallel to transcendentalism

The Object, as distinct from the Principles, gives a general picture of the aim of the Socialist Party of Great Britain. This consists in the creation of a new and radically different society which will overthrow the present capitalist system in this country, and state capitalism in Russia, China and certain European nations. The new society will be based on common ownership and democratic control of the means of production, which will be in the interests of the whole community.

The Principles state more precisely the guiding characteristics of members. The master class (of capitalists) enslave the working class with the consequence that a class struggle is manifested between those who produce and those who own the means of production. The social evolution of mankind implies the inevitable emancipation of the working class, who in the total evolutionary process will be the last identifiable group to achieve this freedom, which will emanate from the working class itself. The government and armed forces, representing the powers of capitalism, must be overcome by peaceful means, oppression will become emancipation. Freedom, equality and comfort will be manifested by the overthrow of capitalist oppressors through the use of <u>peaceful</u> electoral procedures, and because all parties are in reality merely the expression of class interests the SPGB is consequently hostile to all existing political parties.

In many respects, therefore, the SPGB is an avowedly secular body exhibiting many of the characteristics of a transcendental group. This is demonstrated, firstly, by its historical separation from the mainstream of socialist activity. Secondly, there is hope of a millennium in the non-too-distant future, during which time a truly socialist society will be in existence, and all those "evil" associations of capitalism such as monetary exchange, the wages system, and so on, will be overcome. Thirdly, a distinctive twain cosmogony is apparent in the designation of humanity into "the children of socialism and of darkness". Not because those of darkness are necessarily evil but because they have not seen the light of socialism - as defined by the SPGB's exclusivist standpoint. In the last analysis, when the millennium arrives, the distinction between socialism and non-socialism will disappear.

In more general terms, the ideology of the SPGB, as it is frozen in the Object and Principles, demonstrates transcendental characteristics paralleled by overtly religious groups. For example, there is an eschatological search for total truth; there is an explicit dogmatism contained in both the Object and Principles displaying an almost credal nature; there is a totalism inherent in the claim of socialism to be a complete "world view" that will explain everything; there is an intolerance of others, both right wing elements and those professing to be socialists. The SPGB has the only complete access to a pristine and total truth of socialism. Because they practice a "quiet revolution" there is a lack of fanatical extremism and a complete willingness to be open to public scrutiny. The embodiment of a religious rationale is demonstrated by the vocabulary of the Party, and thus it is both part of, and an aid to, the presentation of a total ideology which isolated and dramatises the great and classic struggle between the forces of darkness and light. It is this total ideology which articulates the dilemma of many Party members and designates the appropriate responses. Not only does the great "world struggle" highlight certain moral crises but it is itself the moral crisis.

The SPGB, like a hundred other political sects, displays many characteristics of both religious sects and a transcendental ideology in general. Both categories are concerned with the management of the identity of their members in relation to a total and embracing ideology, and the content is in many ways less important than the behaviour in small value-orientated groups which is designed to facilitate their task - of socialisation - in some manner. It is for this reason that identity management groups of both categories will be expected to construct structures that are basically similar because there exist no other identifiable means of carrying out or facilitating the task in hand. In other words, there exist a minimum of constancy of interaction patterns and symbol forms that are necessary for groups concerned with "putting over" ideological legitimations, resocialisation procedures and so on. These deviant and total world views locate and "solve" to each participant's satisfaction the moral and social dilemmas of the age. The contents of ideologies, secular or transcendental, embrace the individual, carrying him forward in time and space, away from a senseless world, giving him an identity with others who share his views. These groups are cosmological settings for total ideologies. Because the world is socially constructed our cosmologies, also, are so. They reflect the way we perceive ourselves in relation to the universe, and they freeze or catch into their schemes a thousand facets of existence which we hold as significant, endowing the whole thing with meaning.

Conclusions

We have highlighted some of the characteristics which the SPGB, an avowedly fundamentalist political sect, shares with transcendental sects. Its distinctive "open" style of organisation is established by its premise that its particular message is to be subject to rational scrutiny; a characteristic it shares in common with the Society of the New Church. The identity of its members, like the members of other sects, is immediately recognisable by outsiders because of the nature of the exclusiveness and commitment which it reveals. The goals of the SPGB, like those of other sects, are not immediately obtainable and are acknowledged as such by members. Like other sects, too, it seeks to establish a particular event upon earth, not, as in other cases the Will of God or the Reign of the Saints, but the coming of socialism and the furtherence of the socialist ethic. Such an implementation is not solely in the hands of the SPGB although undoubtedly they are the special heralds of this new dispensation. Once it arrives the SPGB will vanish, its task accomplished, into oblivion. Similarly, the Irvingites recognise that the wheel has come full circle, and recognise in the events of the contemporary world sufficient evidence that their task is accomplished.

Like all sects, schism and dissent has been experienced by the SPGB, notably with the events that led up to the expulsion of Fitzgerald and Hawkins. The ultimate sanction of a voluntary movement is and remains expulsion. Like most sects the SPGB is a world-changer, and by the systematic application of rational criteria it seeks to educate the world to socialism. Its members belong to the socialist brotherhood, the brotherhood of all believers.

Like all sects the SPGB lays exclusive claim on its own distinctive interpretation of socialist truths, and the pristine purity of such truths is sustained in the unchanging nature of the Object and Declaration of Principles. To indicate that change might be possible is to cast considerable doubt on the prior history of the movement.

The organisational requirements are those which clearly fulfill those that are minimal for sect organisation,[50] that is, agencies for convening meetings, decision-making mechanisms, guardians of belief and practice, entry and exit procedures for members, strict tests for entry and determining matters for expulsion, the socialisation of new members, and external liaison officers.

So far, we have argued that the SPGB is a voluntary association which members join after demonstration of some proof that they ascribe to the Object and Declaration. The ideology of the movement is strictly adhered to, in order, primarily, to safeguard self and public image of a moral socialist elite which alone is the possessor of the true socialism, and the members of which have a duty to approximate to. Wary of the "cult of personality" it emphasises a democratic priesthood of believers united in a hostility to all other movements.

Conversion to the SPGB appeared to be mostly of a gradual kind but many members described it as "seeing the light" or as being "the only way":

> Life didn't mean very much before, you know. I had played around with the Labour Party but when I received the SPGB literature it was fantastic. Everything I ever wanted was there. I've never looked back.

Sometimes the descriptions are reminiscent of Koestler's conversion to communism, like "the whole universe falling into pattern like the stray pieces of a jigsaw puzzle assembled by magic at one stroke". The old Adam and the new Man.

Once commitment has been declared it has to be maintained and sustained. This is accomplished by the member being seeped in the ideology of the movement, by attendance at meetings and by regular reading of the literature. Some members spoke proudly of their bound back copies of the Socialist Standard, for example. Unfortunately, the result is a mechanical Marxism consisting of stereotyped responses to the evils of the present social order.

The study of sects, then can be extended outside a Christian tradition to groups which are avowedly secular in nature:

> Groups with situationally transcendent ideals seem to display similar organisational types. This may be seen in political parties with some degree of utopianism in their programmatic aspirations.[51]

Specifically in terms of religious and political sects Park has remarked:

> With respect to the whole-souled devotion which individuals sometimes exhibit to a political faith which they profess, and with respect to the fanaticism with which they sometimes give themselves to a secular cause, there seem to be no difference between members of a religious and political sect.[52]

and he goes on to say:

> Political parties are characteristically accom-

> modating and comprising, but the sect is intransigent and irreconcilable.[53]

and:

> The sect is radical, just because it seeks to formulate, not a new political programme but a new life-policy.

Undeniably the SPGB reflect many of the characteristics of transcendental sects. This has implications not only for this study but also for the study of sects in general by bringing to light a store of writing devoted to the political sect. One of the consequences might well be to contemplate the whole area on a far broader plain, that is, to embrace all change-seeking movements as they relate to society in general.

NOTES TO CHAPTER SEVEN

[1] Leon Trotsky, The Death Agony of Capitalism and the Tasks of the Fourth International, New York: Pioneer Publishers, 1964, p.56.

[2] Scipio Sighele, Psychologie des Sectes, Paris, 1898. English excerpts can be found in R.E. Park and E.W. Burgess, Introduction to the Science of Sociology, Chicago: University of Chicago Press, 1924, pp.202-207.

[3] Kurt Lang and Gladys Land, Collective Dynamics, New York: Thomas Y. Crowell Company, 1961.

[4] Lewis Coser, "Sects and Sectarians", Dissent, Vol.1, No.4., 1954.

[5] Henri Desroche, Jacob and the Angel, Amherst: University of Massachusetts Press, 1973, en passent. See especially Henri Desroche, "Socialism et sociologie du Christianisme", Cahiers Internationaux de Sociologie, XXI, 1956, pp.149-167. Reprinted in translation in Norman Birnbaum and Gertrude Lenzer, Sociology and Religion: a Book of Readings, Englewood Cliffs, New Jersey: Prentice-Hall, Inc., 1969, pp.215-225.

[6] Birnbaum, ibid, p.219.

[7] Ibid, quoted on p.223.

[8] S.M. Lipset, Political Man, London: Mercury, 1963, p.103.

[9] Vatro Murvar, "Messianism in Russia, Religious and Revolutionary", Journal for the Scientific Study of Religion, Vol.10, No.4, 1971, pp.277-338.

[10] Ibid. p.283.

[11] Ibid. pp.284-5.

[12] For example, David Apter, "Political Religions in the New Nations", in Old Societies and New Nations: the Quest for Modernity in Asia and Africa, ed. Clifford Geertz, New York: Free Press, 1963; E.J. Hobsbawn, Primitive Rebels, Manchester: Manchester University Press, 1959; Sylvia Thrupp, ed. Millennial Dreams in Action, The Hague: Mouton, 1962; K.O.L. Burridge, Mambu: A Melanesian Millennium, London: Methuen, 1960; Peter Worsely, The Trumpet Shall Sound, London: MacGibbon and Kee, 2nd Edition, 1968; Peter Lawrence, Road Belong

Cargo, Manchester: Manchester University Press, 1964; Vittorio Lanternari, The Religions of the Oppressed: a Study of Modern Messianic Cults, New York: Knopf, 1963.

[13] Cary Schwartz, Sect Ideology and Social Status, Chicago: University of Chicago Press, 1970, p.12.

[14] Paul Halmos, The Personal Service Society, University of Wales: Cardiff, 1966.

[15] Alisdair Kee, The Way of Transcendence, England, Penguin, 1971, p.214.

[16] Philip Slater, Microcosm, New York: Wiley and Sons, 1966, en passent.

[17] D.G. MacRae, Ideology and Society, London: Heinemann, 1961, p.64.

[17] K. Marx and Frederick Engels, The German Ideology, pts. I and III, ed. R. Pascal, New York: International Publishers, 1939, p.14. Quoted in Ideology, in Dictionary of the Social Sciences, ed. J. Gould.

[19] Schwartz, op.cit., p.3.

[20] Ibid. p.5.

[21] M. Ginsburg, "Facts and Values", Advancement of Science, Vol.XIX, 1962, p.12.

[22] Schwartz, op.cit., p.10.

[23] Ibid. pp.11-12.

[24] Ibid. p.13.

[25] Philip Rieff, The Triumph of the Therapeutic, New York: Harper Torchbooks, 1966. He writes, "a heavily remissive psychotherapy may become a permanent insitutional fixture of modern culture - a kind of secular methodism for those who remain obstinately uncomfortable in their pleasures". p.238. See also Donald Stone, "New Religious Consciousness and Personal Religious Experience", Sociological Analysis, 1978, 39, 2, pp.123-134.

[26] New Living, Elysium Institute of Los Angeles, p.31, quoted in Schwartz.

[27] William Shepherd, "Religion and the Counter-Culture - A New Religiosity", Sociological Inquiry, 42, I, 1972, pp.3-9.

[28] Kenneth Jones, "Secular Love, Secular Transcendence and Unbelief", Modern Churchman, 1974.

[29] Colin Campbell, Towards a Sociology of Irreligion, 1971, London: MacMillan. This gives a description of rationalist movements in the nineteenth century. See also Talcott Parsons and Robert Bellah in Rocco Caporale and Antonio Grumelli, eds., The Culture of Unbelief, London: University of California Press, 1971. See also Thomas Robbins, Dick Anthony and James Richardson, "Theory and Research on Today's New Religions", Sociological Analysis, 1978, 39, 2, pp.95-122.

[30] Edward Shils, Ideology, in International Encyclopedia of the Social Sciences, 1968, p.68.

[31] Ibid. p.72.

[32] G. Thayer, The British Political Fringe, London: Anthony Blond, 1965.

[33] Betty Reid, Ultra-Leftism in Britain, London: Communist Party Booklet.
The SPGB is related to other parties that constitute the extreme left, but it is particularly similar to the old ILP. The latter opted out of the Labour Party in 1932, though until then it had prospered along with the Labour party from its beginnings in 1893. It produced a number of individuals who later were to become well known in the main party. In the 1929 General Election it claimed a total of 37 MPs. After its break with the Labour Party the group has drastically declined.

Like the SPGB it is now a numerically small and powerless party. Like the SPGB it, too, espouses a naive socialism that appears to be unchanged from some sixty years ago, and communicates in hackneyed and stereotyped socialist phrases. If we estimate the membership of the SPGB as 711 then it is little surprise to learn that ILP can boast 250. The SPGB can claim Joan Lister as an ex-member but ILP can refer to Keir Hardie and Fenner Brockway. In terms of goals those of the ILP are unspecific and vague although, unlike the SPGB, they are not resistant to some form of agreement with other socialist groups.

[34] The Socialist Standard, September, 1954, p.1.

[35] Questions of the Day, London: Socialist Party of Great Britain, 1969, pp.66-68.

[36] The Socialist Standard, No.325, Vol.28, September 1931.

[37] The Socialist Standard, 1931, p.2.

[38] Ibid. The place is given as Blackburn, although another official account, in Questions of the Day, gives the place as Burnley.

[39] Ibid. 1931.

[40] Questions of the Day, 1969, p.67.

[41] See Peter Rolling. Studying a Small Powerless Party, 1966, unpublished manuscript. I am very indebted to Peter Rolling for help in supplying information relating to the SPGB, including introducing me to their head office.

[42] Rolling, op.cit. pp.10-12.

[43] Ibid. p.14.
Rolling's work, although never published, is still as accurate today, as it was seventeen years ago, and my own work bears this out. If we take his six "consistencies" we still find the same basic stance as was evident in the very beginning, and the same fundamentalist approach to the "D of P". Prospective members, however few in number these might be, are still screened on entry by being given a series of written questions. The fundamentalist nature of the Socialist Standard is still very much a reality, and members constantly refer to it with marked reverence. This consistency of output is certainly effective in achieving a coherent and shared ideology. Members still insist on a hundred per cent adoption of their "D of P", and exhibit the same consistent hostility to other political parties of both the Right and the Left as they have always done. Known as the "Small Party of Good Boys", the SPGB practices a quiet revolution primarily through education. They make no secret of their records or procedures and are open to public scrutiny. They are friendly and welcoming - even to researchers!

[44] Ibid. p.8.

[45] The Socialist Standard, February, 1970, pp.29-30.

[46] Western Socialist, Vol.36, No.267, 1969, p.19.

[47] Western Socialist, Vol.28, No.228, 1962, p.13.

[48] Peter Lawrence, "Towards One World", No.3, Aspects of Religious Socialism. Mimeographed.

[49] This characteristic is typical of many transcendental and secular groups exhibiting a total ideology, as typified in Leon Festinger's When Prophecy Fails.

[50] Bryan Wilson, ed. *Patterns of Sectarianism*, London: Heinemann, p.14.

[51] Thomas O'Dea, "Sects and Cults", in *International Encyclopedia of the Social Sciences*, New York: MacMillan and Co., 1968, Vol.14, p.135.

[52] R.E. Park, "Characteristics of the Sect", *On Social Control and Collective Behaviour*, selected papers edited by R. Turner, Chicago: University of Chicago Press, 1967, pp.240-248.

[53] Ibid. pp.245-246.

[54] Ibid. p.245.

8 The organisation and structure of two secular value-oriented groups: *Recovery, Inc.* and *Neurotics Nomine*

Transcendental groups, we contend, often display an organisation and structure either very similar to secular groups or in many ways difficult to separate. We have seen how Alcoholics Anonymous, for example, exhibits quite clearly patterns of ideology and organisation which are transcendental or sect-like.[1] Various studies of humanist societies have again suggested that these groups are very reminiscent of religious sects.[2] Studies of overtly secular groups such as the Socialist Party of Great Britain, the Internationalists and other fringe political movements highlight them as possessing many of the characteristics of transcendental groups.

We have suggested that others have indicated a similar pattern to be found in therapeutic and revolutionary groups. For example, Schwartz sees convincing "parallels between the development of chialistic and apocalyptic religious visions in earlier phases of Western history and contemporary revolutionary doctrines of both the Right and the Left".[3] This observation is not limited simply to political ideologies. Rieff's[4] work on the schismatic followers of Freud, and Slater's[5] analysis of analytic groups, further suggests that an analysis of sect is not to be confined to the transcendental realm. Some time ago Berger[6] suggested that "the dynamics of sectarianism" is to be found "at work in places far removed from religion proper - in politics, art, literature, and even within the sacred precincts of science itself".

Recovery, Inc. and Neurotics Nomine are both movements which reveal similarities to religious sectarianism both in their belief structure and organisation. An examination of some of the characteristics of resemblance will, it is hoped, throw some light both on these groups themselves and also confirm or modify our perception of the way in which sects are structured. The two movements are commonly referred to as "ex-patient" organisations, that is, groups specifically formed for members to share common problems and experiences arising out of their life-situations as former mental patients.[7]

Growth and development

Wechsler attributed the growth of ex-patient organisations to the "other-directed" and "organisation-orientated" drift in modern society in which the mentally ill are becoming joiners as much as any other section in the population.[8] There are naturally important differences in the style of participation. In the community as a whole, and especially in the United States, voluntary association membership can be as high as sixty-five to eighty per cent. But as others[9] point out often as many as sixty-five per cent drop out suggesting that an individual's membership profile varies considerably even in a short period of time. Many have multiple memberships. Affiliation is most common in the middle years and is linked to family status and life cycle. After the age of seventy there is a decline in affiliation. Generally speaking voluntary associations fall into two main categories[10] which have been termed social influence and expressive groups, the former endeavouring to achieve a condition of change in some aspect of society and the latter formed to express or satisfy the interest of their members. There is some suggestion, notably by Chapin,[11] that voluntary associations develop into bureaucratic structures over time. Gordon and Babchuck frame a typology incorporating what they regarded as the main factors involved in voluntary affiliation, that is a) degree of accessibility; b) status conferring capacity; c) function.

In contrast to these voluntary associations belonging to the general community the particular problems and status of ex-mental patients make it difficult for them to affiliate easily to most voluntary organisations and enhance the need for affiliative groupings of a more suitable order. Ex-patients find it particularly difficult to sustain social relationships and to encounter new ones. The whole problem of the stigma[12] attached to mental illness forces ex-mental patients to create their own stress free situations operating on the principle of "like to like". This is the major motivating factor given for the emergence and perpetuation of these therapeutic groups.[13]

Studies of types of ex-patient organisations

Wechsler ranged ex-patient organisations on a continuum based on the way they relate to the "supportative milieu", [14] and he categorised twenty per cent as being patient-led as compared with eighty per cent which are professionally-led. "Professionally-run groups are usually sponsored by state agencies, hospitals or mental health associations, while patient-run groups must obtain their own funds and facilities". Examples of the former include Fountain House, San Francisco Fellowship Club, and Center Club, and the latter Recovery, Inc., Search and Club 103. Some of these are in fact halfway houses, that is "a residential institution designed with the ostensible goal of affecting the transition of the ex-patient from the mental hospital to the community".[15] Wechsler sees the choice as a reflection of the particular ideology which is held. The professional-based group views the patient as not yet ready to assume the responsibilities inherent in running such an organisation. The patient-based group, on the other hand, is often seen as promoting initiative and independence. In practice, of course, the two extremes often come together. A further differentiation of type noted by Wechsler is between groups which base themselves on a "sociability" function and those which employ other methods of treatment.[16]

Membership of ex-patient organisations throughout the world is probably not more than some twenty thousand. Most organisations "are relatively small in size, with two thirds having twenty or fewer regularly attending members".[17] Groups themselves differ over membership policy. "Alumni" organisations are open only to ex-patients from specific hospitals, while "total" or "inclusive" organisations permit anyone who has been treated for illness to become a member. Some, like Recovery, Inc., have some members who have never been hospitalised. Professionally run groups run a screening procedure, and aspirants who are alcoholics, psychopaths or sexual deviants are dissuaded from joining, generally on the grounds that they will disrupt other members.

Activities of groups vary enormously. Some are essentially social rehabilitation centres[18] while others place an emphasis on a "re-education" of members. Visits, indoor games, talks, and so on, are all part of the recreational pattern that these clubs attempt to provide. In theory it is felt that such activities offer a rehearsal procedure that will facilitate entry into the community.

Lay-run groups have a shorter life-span[19] than professionally run groups due not only to the difficulty in obtaining members but also to conflict in self-governing policies. Recruitment difficulties mainly reside in the individual having to assume the "ex-patient role", which makes "passing" in the community more difficult. Many patients wish to avoid the "labelling" which would accompany affiliation to these clubs.[20] In addition, lay-run groups are far less formally and efficiently organised than professional ones. Goertzel, Beard and Pilnick's study of the Fountain House Foundation, one professionally organised ex-patient group, reveals a highly formalised organisation with a thirty-eight member

Board of Directors and thirteen Standing Committees.[21] In 1960 it had an annual budget of over $200,000 and a total active membership of five hundred and fifty.

The members were a mixture of schizophrenics, psychotics and non-psychotics. "Men outnumber women three to two. Only five per cent are twenty years or younger and only eight per cent are over fifty years of age. Two thirds of the members are in their twenties and thirties. Only one out of twenty is currently married."[22] Looked after by a full-time professional staff of thirteen this organisation sponsors a transitional employment programme and a social recreation programme. The former includes placement at various occupations, as "trys", and the latter an evening and weekend series of organised activities. This kind of formal organisation stands in marked contrast to many of the lay-run bodies, and indeed to some other professional ones.

Landy and Singer[23] were primarily concerned with the way the clubs for former mental patients were socially organised, and with their general culture. Basing their observations mainly on Club 103 they make the interesting point that the club movement may in time become a kind of cult involving highly ritualised and formalised behaviour with a "charismatic" leader who expounds the Word. Such cults have their dogma and exegesis. Landy and Singer instance Recovery, Inc., as a movement in which the "Book", Mental Health Through Will Training, has become a "bible", a point to which we will later return. In discussing organisational development Club 103 had no formalised rituals of admission. There was a hard core of daily attenders (fifteen frequent and twelve regular) who met for about six hours a day. The "older" (longest and most frequent) became de facto representatives on the club-hospital steering committee. There was the emergence of an informal leader, but this did not last for long. Basing their analysis on Redfield's "Folk-culture" theme,[24] they showed that the club offered both positive and negative benefits. For example, not only did members share a common culture to "feel out" their hospital experiences but they saw in the club a community which offered freedom from the demands of the outside world, e.g., "dating" and social ranking. The social organisation of the club consisted of a small inner core and a larger peripheral membership. There was a loose structuring of social relationships and an apparent need for leadership and fellowship. Social relations were "equal" but separate, with a general levelling of social status. The basis of relationships was socio-emotional, and in the sense that it was a self-perpetuating group for the perenially sick it resembled an informally organised deviant sub-group.

Rapoport[25] describes how the community meetings of one group, the "8-30", resembles group therapy sessions, with the exception that there are some one hundred and twenty five members rather than a dozen. The group has an intimate self-confessional aspect that he likens to that found in the congregations of religious sects such as Buchmanites and Quakers that stress leaderless public confessionals. It is not viewed as primarily a treatment group although treatment does undoubtedly occur, and there are powerful forces at the disposal of the treatment aims of such a group, principally social control. The devices used are varied and include: data gathering, "defaulters" meeting, entertainment committee, and a communication channel (verbalisation problems stemming from socio-economic class). The groups are further organised into ward meetings, departure groups, family groups, socials, doctors' groups, and leisure groups. At a "defaulter's" meeting, for example, the culprit's misdemeanour would be highlighted by the group and steps taken to bring him in line.

Sugarman[26] described in detail the activities of "Daytop", a community therapy institution for ex-criminals and addicts. Daytop displays a strong hierarchy of leadership. On admittance senior residents tell the addict that he is <u>responsible</u> for his condition and no one else (a form of "reality therapy"). In effect the addict is receiving a ritualised "dressing down" after which, gradually, by a seemingly harsh programme, he begins to <u>relate</u> to his fellows. The morning meeting is highly ritualised and includes a time devoted to formal "pull-ups". After the "pull-ups" there is a series of announcements. Groups in Daytop are <u>static</u> (the same people meet each time) or <u>encounter</u> (based on grievances residents might entertain against each other). He sees them as providing two functions - a) offering controlled release of hostility and verbal aggression; b) forming a setting for "reality therapy". In his description of the Phoenix Unit[27] Sugarman stresses the importance placed on the patient's relationships with other inmates and non-medical staff. The Phoenix Unit is a heterogeneous therapeutic community serving sixty-five "in" and twenty-five "out" patients, with an ideology which views illness as intimately bound up with interpersonal relationships and an inability to cope with deepseated emotional feelings. Consequently, an expression of such feelings becomes an important aspect of the Phoenix treatment.

A community similar to Daytop, Synanon, is described by Volkman and Cressey.[28] Basically it is run on the same lines. It does not claim to "cure" drug addicts but merely to help them stay away from drugs. There is a hierarchical movement up through the "stages" from new entrants to "graduates" which reflects the success of the passage through the movement, and representing "stages of graded competence".[29]

Recovery, Inc.: The association of nervous and former mental patients

Recovery, Inc., is the largest and probably the oldest existing self-help group for former mental patients with over six thousand regular paid members and seven hundred and twenty-five groups in forty-one states in America. Little has been written about it and subsequently most of the information is to be gleaned from the "Book" of the movement.

Mental Health Through Will Training[30]

The movement was founded on November 7th, 1937, by thirty patients, or rather ex-patients, who had previously been treated at the Psychiatric Institute of the University of Illinois. Between 1937 and 1940 the organistion was limited to patients admitted to wards of the Psychiatric Institute, but in 1940 it extended its work to out-patients in that establishment. In September, 1941, the group broke away from the University and established its own headquarters in 116 South Michigan Avenue. In the years after 1942 the bulk of the membership was recruited from the private practice of the main founder and writer of the Recovery Book, Dr. Abraham Low.[31]

Recovery, Inc. offers group psychotherapy classes prior to patients leaving hospital in order that they might better face the problem of community living. The organisation also offers a complete programme of recovery during both the "phasing-out" period and also for those who have left the psychiatric institution. Members are protracted chronics, that is their range of suffering is from two to twenty years. Panel members' groups or listeners' groups take place some three times a week during which time, generally speaking, members discuss a chapter or an article from one of Low's books, or from Recovery Journal or Recovery News (the News having succeeded the Journal). Group psychotherapy classes are conducted by a physician. The Public Meeting is "open" and consists of a panel discussion and the physician's address. Panel meetings are led by a Panel leader. Throughout, the appearance of "setbacks" is catered for by veteran Recovery members being at hand to be contacted in distress. If the veteran fails to accomplish anything the Panel leader is called, and if that fails the Chairman of the organisation is called. As a last resort the physician is called, although this rarely occurs.[32]

Recovery Language

The symptomatic idiom has to be learned by members, including facial expression and gestures.[33] The "Recovery Language"[34] includes conceptually important phrases such as "sabotage" and "authority". The authority of the physician is "sabotaged" if a member presumes to make a diagnostic, therapeutic or prognostic statement. Once a physician has diagnosed a member's condition the latter is dissuaded from "self-diagnostic revelry". The "spotting technique"[35] encourages the member to "spot" inconsistencies and fallacies in his own language in order to check his sabotaging propensities. "Spotting" brings home to the member the fact that there is no danger involved in a situation. One of the functions of the Recovery Language is that it enables the member to bring a situation within his control.

Development

Since the early days[36] social events and activities were quite common but Low later tended to play these down. In the beginning they consisted in what were tantamount to personal chats on an informal basis with refreshments, or Kaffeeklatsch as Low called them. In 1950[37] the membership fee was $3 per year, and since then the movement has been open to any psycho-neurotic or former mental patient. In January, 1950, the membership was three hundred and seventy-six, seventy-five per cent of whom were procured from Low's private practice. The movement largely finances itself by private donations, fees, sale of Recovery literatures and collections at meetings, although in the early days Low met the yearly deficit. The affairs of the organisation[38] are now conducted by a Board of three directors, all ex-patients. There is also a Vice-President, a secretary, treasurer and six counsellors, although Low himself never held office.

Literature

The editor of Recovery News,[39] which after the eleventh issue replaced Recovery Journal which ran from June, 1946 to June, 1947 and collapsed owing to publishing costs, is a former mental patient. The publication itself appears eight times a year. Recovery literature between 1938 and 1941 consisted of a bi-monthly magazine Lost and Found edited by Low who also contributed most of the articles. In 1943 this material was issued in book form in three volumes The Techniques of Self-Help in Psychiatric After-Care which has now sold several thousand copies. The first edition of Mental Health Through Will Training appeared in 1950 and the sixteenth edition in 1968. The book is essentially a series of case studies designed to illustrate examples of post-neurotic symptoms and a proposed aid to recovery.

Recovery, Inc., controls the publications of the movement, a pattern shared by such sects as the Swedenborgians, the Catholic Apostolics or Irvingites, the Exclusive Brethren and the Witnesses, as well as many others. [40] To some extent, of course, control of the published information means in effect control of the ideology of the movement. With the Christadelphians the editor of the official journal became, effectively, leader of the sect, and in 1917 the directors of Christian Science took action in the courts to gain control of the publishing enterprise in order to consolidate their leadership. Such an analysis is not to be confined solely to religious groups. Alcoholics Anonymous and the Socialist Party of Great Britain have a similar value placed on their publications. Such publications not only determine to a large degree the external image of the movement but they become in turn a source of social control in groups lacking formal leadership roles. For example, in Recovery, Inc., a literalistic adherence to the Book is both encouraged and in turn utilised as the official yardstick for the group's ideology.

Later developments

Since the original publication of the Book, Low made several changes. However, it was not until after his death in 1954 that the organisation assumed a different and independent character. Now Recovery meetings are open to the public; social activities are no longer part of the Recovery method - in fact Low, just before his death, asked that these aspects be played down. There is now a maximum amount of training in self-help activities.

During its formative years, Recovery, Inc., was a professionally supervised organisation. The initial aims were gradually to: help incidences of relapse in mental illness and prevent a chronic state developing; save time for the physician and expense for the patient; and to create an independent self-help lay group without professional supervision. In 1952 Low established the Panel example method (three or four experienced members discuss a chapter from Low's three volumes on self-directed after care) which became a model throughout the entire organisation. Panel leaders have received: a) authorisation by Recovery's Board of Directors; b) a thorough training (at least one week!) in the method. Each group contains some thirty members, and a trained leader is responsible for procedure. Referral is through doctors, psychiatrists, mental hospitals and clinics. After Low's death in 1954, the organisation withstood the death of the founder and emerged as an independent self-help, lay organisation. Because of its lay character it does not replace the physician whose authority remains constantly paramount, and neither does it purport to offer diagnosis, medical treatment or counselling. It is non-sectarian, non-political, and has no affiliation with any other organisation or individual. It is non-profit making, and utilises psychotherapeutic methods to deal mainly with post-psychotic and psychoneurotic persons.

Wechsler's case study

Wechsler distributed one thousand eight hundred and seventy-five questionnaires and recived seven hundred and seventy-nine in return. The model Recovery member who emerged was middle-class, middle-aged, married and female.

"She has had at least some high school education, and may have attended college. Her husband is employed in a non-manual occupation and the yearly income is approximately $6,000."(1960) He goes on to describe the profile as having one or more independent children. "She is active in community and church affairs, belongs to one or more voluntary community associations and attends church services weekly. She takes part in various leisure time activities...Thus she is an individual who appears to be integrated into her community and who, at least on the basis of certain socio-economic criteria, appears to hold an average or perhaps better-than-average status." [41] One hundred and twelve of the returned questionnaires were from leaders of Recovery groups, who approximated to the general profile, with the exception that they were of a higher socio-economic status and displayed more voluntary activity. Wechsler cites three reasons for joining Recovery. First the patient suffered from psychological symptoms such as fears, delusions and "nerves". Second the patient suffered from psychosomatic symptoms, such as tremors and heart palpitations. Third there was a genuine curiosity as to whether the organisation could help.[42]

The Recovery method seeks to alleviate the fears of the patient that psychoneurotic or postpsychotic symptoms are dangerous, and part of the method's success lies in the patient accepting the diagnosis of the doctor and subsequently applying his "will" to alleviate threatening symptoms. The Recovery principles include "spotting", the Recovery language, and so on, but furthermore the patient is taught to distinguish between the external environment, which cannot be changed, and his own internal environment, which is subjective. The patient is to refrain where possible from a commitment to judgements of "right" and "wrong", and to consider himself at all times as average rather than exceptional.

Wechsler's account of Recovery meetings illuminates the organisation and structure of these groups. The meetings are strictly routinised or liturgical and are attended by between eleven and thirty members. According to him such meetings are divided into four parts. The first part is devoted to readings from the Book, which has the dual effect of therapeutically engaging the participant more actively and also familiarising him with the contents of Low's method. The second part is concerned with the presentation of the examples for which there exist procedures which are strictly adhered to. Various "models" are incorporated in Low's book which exemplify the procedures. The examples can be characterised by: a) familiarity - the context is always that of the familiar; b) concreteness - specificity of illustration; c) somatisation of symptoms - a general symptomisation expressed most frequently as air hunger, heart palpitations, perspiration and tremours; d) uniformity of symptoms - the same symptoms are referred to; e) exclusive use of Recovery concepts - extraneous terms are excluded and the Recovery terminology adhered to; f) the testimonial - the success of the method is demonstrated by its use. The third part comprises group participation during which the principles and techniques of the Recovery method are utilised. The fourth and last part is a question and answer period which is usually no longer than fifteen minutes. After all this refreshments are served which become "mutual aid" periods.[43]

Wechsler particularly draws out the cathartic value of the Recovery confessional[44] which, he suggests, may well "aid the individual in removing some of the anxiety associated with his symptoms". The act of public admission reinforces in the patient the knowledge of the similarity his symptom has to others. In most respects this is similar to the confessional period of Alcoholics Anonymous and many religious sects. Wechsler also stresses the religious or semi-religious nature of Recovery as a total programme. "The Recovery method and its practice at Panel meetings is clearly reminiscent of various elementary characteristics of certain organised religions. The method involves faith and acceptance of regulations handed down by a higher authority. The method stresses self-discipline and the volitional aspects of human nature. This type of approach is more amenable to certain religious beliefs than is the Freudian and Darwinian notions of man as an instinct-driven animal. The emphasis on the power of positive thinking and on inspiration is also clearly analagous to some religious tenets."

He goes on to add that "the analogy between Recovery and religion may be extended further. Recovery has a bible, the textbook of Dr. Low. Hero worship of Low sometimes assumes the proportions of making him almost appear a god-figure. The leaders assume the role of disciples. In certain Recovery groups the desire for expansion and for national recognition is analogous to the missionary zeal in religious groups. In addition, the repetitive ritual-like Panel meetings resemble certain forms of religious ceremonies".[45]

The group leader in Recovery, Inc., "assumes the role of a nervous patient". He must appear expert in the nuances of this specific methodology. He can also, however, curtail a critical discussion of the method of directing the group onto something else. In this way the exercise of the role is also the exercise of a powerful means of social control which ensures that members retain their fervour and adherence to the aims of the movement.

Neurotics Nomine

Neurotics Nomine is a voluntary organisation under the

presidency of Dr. Joshua Bierer. It was established some twenty four years ago to meet the increasing anxiety felt by some at the rise of the numbers of people with nervous or mental illness. It was thought that such illness prevented a successful adjustment to normal activity within the community, and that such an organisation as Neurotics Nomine could provide a supportive role in bridging the gulf that existed. A precendent had already been established by the formation of several other self-help groups such as Alcoholics Anonymous and societies for the blind, hard of hearing, crippled, and so on. In other words, there have been several secular responses to what Glock and Stark called "organismic" deprivation. [46] "Helping groups" were formed by ex-mental patients and others interested in nervous and mental illness. Neurotics Nomine benefits from the advice of the psychiatrist but is lay-run and controlled essentially by ex-patients and patients who are still in the process of receiving treatment, although community-integrated to some extent. Like Alcoholics Anonymous, Neurotics Nomine stresses anonymity although outside the organisation this is left to individual choice. It is now a registered charity with a membership of some five hundred.

The aims of the organisation are as follows: a) to help ex-mental and mental patients to meet each other and people outside the organisation; b) to assist the patient in taking the first steps towards social adjustment; c) to support each other by meetings at home or talking over the telephone in an emergency; d) to provide a panel available for discussion or to answer questions to an audience of GPs or members of the Social Services; e) to help remove the stigma and fear that still surrounds mental illness; f) to help the members by providing, through these activities, opportunities for useful work, social contacts and self-expressiion.

The organisation and its environment

The aims listed above involve us in a complex problem which has constantly confronted sociologists of organisations. It is not at all clear, for example, that the goals of formal organistions are as clear-cut as some writers have implied.[47] Can we, in fact, make logical sense when we attribute goals to educational organisations? Or are we in fact simply falling into the trap of reification?[48] These sorts of problems involved in the study of organisations require separate treatment. Suffice it to say here that the stated goals or aims of Neurotics Nomine may, as the organisation develops, become displaced, original goals may make way for current goals, and observers of such organisations may infer quite different goals from those inferred by members.[49]

"Helping groups" such as Neurotics Nomine may have a certain ambiguity in relation to their environment simply because membership is not at variance with that environment but is undertaken in order to function adequately within that environment. The stigmatisation which mental illness brings to those who function in the normal environment creates a gulf between the individual and that environment. Neurotics Nomine attempts to bridge that gulf by a series of interpersonal activities. members may visit exhibitions and concerts, speakers from "outside" may be brought in, and so on.

Division of labour and communication

Neurotics Nomine (Neurotics by Name) is a lay-run ex-patient organisation with some official backing, i.e., some psychiatrists take a strong interest and sit on committees, as do some full-time Council of Social Service employees. In areas where official backing is weak or non-existent then groups tend to oscillate. At group meetings between fifteen and twenty persons attend on average. Activities in these groups generally consist of informal activities such as bingo-playing, table tennis and the occasional dance. This set of rudimentary activities is designed to draw members into a form of social interaction. Illness is rarely if ever discussed. The principal organiser of suh group meetings is usually an ex-patient who calls everyone by their first name, and there is no professional attendance at these meetings.

The main means of both formal and informal communication is through infrequent issue of _Contact_, the Neurotics Nomine magazine which is usually cyclostyled. The contents consist of reminiscences of travel abroad, poetry composed by members, and "confessional" accounts of personal illness, often accompanied by religious overtones. Other matters contained in _Contact_ are concerned with the debate as to whether the name of the organisation is a suitable one, and whether more personal help could be given by the lay members, e.g., someone to telephone when in distress, or to write to. Some branches have their own newsletter and news of branches is reported in the national magazine.

Systems of engagement and reward[50]

In the case of ex-mental patients the need for a gradual re-entry process into the normal activities of the environment are important, and this is particularly so in view of the process of overt and covert depersonalisation which exists in many of our mental hospitals. The stigmatisation of the mentally ill is still a reality in contemporary society, and these social clubs assist to some extent in the breaking down of this stigma. Further, such ex-patient organisations form a kind of _rite de passage_ for the patient to gain entry into the community and re-establish satisfactory relationships, which become vitally important for the perpetuation of the recovery process. These groups present a gradual induction for the re-establishment process, and they are self-supporting in the sense that most members have been stigmatised in roughly the same manner and consequently can be assumed to understand each other's problems. There is no compulsion to join in the activities of the groups, and individuals may enter into an engagement whenever it suits them. The environment provided by the group is, perhaps, a sheltered and artificial one, but to some extent this is necessary when dealing with so susceptible a membership. The ultimate goal is complete resocialisation into the community.

The demarcation suggested by Wechsler of lay and professional therapeutic groups is not really as clear as he suggests. Neurotics Nomine, for example, has professionals on its committees and many professionally run organisations can have such mechanisms as ex-patients on their committees. Another difference is that some of these groups, as we have suggested, are purely social in function while others have a psychotherapeutic function. Some groups are open only to ex-patients while others admit people who are still hospitalised on "release evenings". Some are completely closed and admit no outsiders while others welcome the community "normals" who profess sympathy with the aims of the organisation. Neurotics Nomine no longer confines itself to neurotics or people with a history of psychiatric illness but extends its membership to the lonely, the depressed and the obese.

We have already suggested that some ex-patients may not wish to identify themselves formally with an ex-patient role seeking rather to operate a system of "passing" in the normal community. Consequently some would argue that belonging to an ex-patient group may well outweigh the advantages. For example, affiliation may well serve as a means of differentiating from the wider community - members may regard themselves as ex-patients rather than as husbands, teachers, business executives or clerks who have been ill - and such clubs may operate manifest or latent mechanisms of reinforcement to this end. Some have been attending groups for over six years and have neither attempted nor wished to allow their membership to lapse.

Organisational structure of the two groups

Neurotics Nomine lacks rigid control and centralised authority. Recovery, Inc., on the other hand, has a more obvious structure and local groups must adhere to a formalised procedure. Neurotics Nomine are autonomous groups[51] run by a voluntary secretary and a committee comprised of a relatively flexible membership.[52] The Executive Committee of Neurotics Nomine is composed of ex-patients. Usually two people are assigned to each task in order to ensure that continuity is maintained. The most successful Neurotics Nomine groups are to be found in areas where there is "good" official backing, i.e., London, Liverpool and Norwich.[53]

Both the groups we have discussed exhibit sectarian characteristics of organisation and structure. They are small value-orientated groups possessing a voluntary membership which is conditional upon some mark of merit (e.g., the overt admission of mental illness, alcoholism, and so on). Membership is nearly always voluntary with the exception of some day therapeutic clubs at which attendance is part of the professional treatment.[55] Although ostensibly the main goal of these groups is to enable the individual to obtain integration into the community nevertheless one of the criticisms levelled against organisations such as these is that they inevitably display separatist tendencies to society in general and withdrawal from the world and its institutions and values. There is a substantive element of self-perpetuation latent in organisations which specifically meet in order that individuals with similar problems may discuss their common ailments. Wechsler maintains that "when no formal mechanism" exists for the termination of ex-patienthood, individuals may be limited in growing out of it. Here, then, these organisations display a characteristic of sectarianism, that of withdrawal and separation.

Recovery, Inc., displays some aspects of conversion activity common to value-orientated movements. The individual must display faith in the Recovery Programme and acceptance of the general regulatory pattern displayed by the organisation. Confessional characteristics are shown not only in the group meetings but also in the overall acceptance of the movement's tenets and the pragmatic aspect of the Testimonial.

Both transcendental and secular movements of this kind exhibit a "totalitarian ideology" which manages the individual in his dealings with the "normal" community.[56] It is in the meetings that the member is taught, either formally or informally, a set of techniques for coping with the day to day exigencies of living. These techniques can consist of either a set of instructions such as those Recovery, Inc. offers or the possession of a certain singlemindedness and fervour which will provide combatant stances with which to cope with his problems. The formal structure of Recovery, Inc. is certainly much tighter than that of Neurotics Nomine, although both offer a structured system of goals, methods and taboos. The former is organisationally geared to the achievement of its goals in a way that Neurotics Nomine is not, although nevertheless the latter is geared in an <u>informal</u> way to such an achievement.

Recovery, Inc. places significant reliance on the Writings of Low which are regarded almost as "new" knowledge received from outside the everyday world. The movement, by use of Panel Leaders, sustains its concern for the purity of these Writings. The ideology of the movement is consequently based on Low's work and any deviation is used to crystallise the orthodox stance by being utilised as an example of "straying" or "backsliding" which serves to consolidate the main position. Members of both groups exhibit a totality of concern in which accommodation to the "normal" society is the primary goal.

It has been argued that similar patterns of organisation and structure are to be found in both secular and transcendental movements. The therapeutic groups under discussion are overtly secular organisations which only inadvertently display "religious trappings". Nevertheless, although the identification is not necessarily complete, secular movements do resemble religious groups. Some of the points of similarity have been suggested. Others could include the suggestion that both types of groups are concerned with the management or resocialisation of the identity of their members in some way. In other words, to do the job required, whether it be the legitimation of an ideology or a form of resocialisation, these groups offer a locus for solving both personal and social dilemmas which confront us. They are not functional alternatives to one another nor to religious sects but rather fulfill the same "competences", to quote Nadel, which are located in society to legitimate both private and public cosmologies.

NOTES TO CHAPTER EIGHT

[1] See Chapter Four and R.K. Jones, "Sectarian Characteristics of Alcoholics Anonymous", Sociology, Vol.4, No.2, May, 1970, pp.181-195.

Ruitenbek, in a discussion of the new group therapies, has argued that "The differences between the revival meetings and marathon group therepy, the Living Theater, a weekend at one of the news institutes such as Esalen, and other forms of therapeutic encounter are largely limited to ideology; otherwise, the quality of human experience is markedly similar. As a matter of fact, the religious revival movements in the United States probably served for many years as a means through which many psychological pressures could be relieved. The simplicity of the revival movement...is probably still very much with us." (Herdrikon Ruitenbek, The New Group Therapies, New York. Discuss Books, 1970.)

Others have argued that a lack of gemeinschaft groups and an increase of the kind of sophistication felt to be lacking in many religious groups, could be a reason for the growth in popularity of a form of new communitarianism or "non-Christian" intensive group experiences. (John H. Marx, Joseph E. Seldin, "Crossroads of Crisis II Organisational and Ideological Models for Contemporary Quasi-Therapeutic Communes", Journal of Health and Social Behaviour, Vol.14, No.2, June, 1973.

[2] Susan Budd, "The Humanist Societies: the Consequences of a Diffuse Belief System", in B.R. Wilson (ed.), Patterns of Sectariansim, London: Heinemann, 1967, pp.377-406. See also Colin Campbell, Towards a Sociology of Irreligion, 1971, London: MacMillan.

[3] Gary Schwartz, Sect Ideologies and Social Status, Chicago: University of Chicago Press, 1970, p.10.

[4] Philip Rieff, The Triumph of the Therapeutic, New York: Harper Torchbooks, 1966.

[5] Philip Slater, Microcosm, New York: Wiley and Sons, 1966, en passent.

[6] Peter Berger, "The Sociological Study of Sectarianism", Social Research, Vol.21, No.4, Winter 1954.

[7] Henry Wechsler, "The Ex-Patient Organisation: a Survey", Journal of Social Issues, Vol.XVI, No.2, pp.47-53, 1960.

[8] Henry Wechsler, op.cit., p.47.

[9] Nicholas Babchuck and Alan Booth, "Voluntary Association Membership: a Longitudinal Analysis", American Sociological Review, Vol.34, No.1, 1969, pp.31-45.

[10] C.W. Gordon, and Nicholas Babchuk, "A Typology of Voluntary Associations", American Sociological Review, Vol.24, No.1, pp.22-29, 1959.

[11] F. Stuart Chapin and John E. Tsouderos, "The Formalisation Process in Voluntary Organisations", Social Forces, 34, 1956.

[12] Erving Goffman, Stigma: Notes on the Management of Spoiled Identity, London: Pelican, 1968.

[13] Wechsler, op.cit., p.48.

[14] Ibid. pp.48-9.

[15] Harold L. Raush with Charlotte L. Raush, The Halfway House Movement: a Search for Sanity, New York: Appleton Centruy Crofts, 1968, p.189.

[16] Wechsler, op.cit., p.50.

[17] Ibid.

[18] Ibid.

[19] Ibid. p.52.

[20] Ibid. p.53.

[21] Victor Goertzel, John H. Beard and Saul Pilnick, "Fountain House Foundations: a Case Study of an Ex-patient's Club", Journal of Social Issues, Vol. XVI, No.2, pp.54-61, 1960.

[22] Ibid. p.57.

[23] David Landy and S. Inger, "The Social Organisation and Culture of a Club for Former Mental Patients", Human Relations, XIV, 1961.

[24] Robert Redfield, The Folk Culture of Yucatan, Chicago: University of Chicago Press, 1941.

[25] Robert Rapoport, The Community as Doctor, London: Tavistock, 1962.

[26] Barry N. Sugarman, "Daytop Village" (unpublished manuscript), 1968. Also "Daytop Village: a Drug-Care Co-operative", New Society, April, 13, 1967, pp.526-529.

[27] Barry N. Sugarman, "The Phoenix Unit: Alliance Against Illness", New Society, June 6, 1968, pp.830-833.

[28] Rita Volkman and Donald R. Cressey, "Differential Association and the Rehabilitation of Drug Addicts". American Journal of Sociology, Vol.60, No.2, pp.129-142. See also Lewis Yablonsky, "The Anti-Criminal Society: Synanon", Federal Probation, XXVI, 1962, pp.50-57, and also Synanon: the Tunnel Back, Baltimore: Penguin Books, 1965.

[29] Other relevant literature on therapeutic communities and ex-patient organisations include: M.B. Palmer, "Social Rehabilitation for Mental Patients", Mental Hygiene, 42, 1958, pp.24-28; D.A.S. Blair, "The Social Therapeutic Club: an Important Measure of Social Rehabilitation in the Treatment of Psychiatric Cases", Mental Hygiene, 39, 1955, pp.54-62; Joseph A. Shelly, and Alexander Bassin, "Daytop Lodge - A New Treatment Approach for Drug Addicts", Corrective Psychiatry, XI, July, 1965, pp.186-195.

[30] Abraham A. Low, Mental Health Through Will Training, Boston: Christopher Publishing House, 1968.

[31] Ibid. p.16.

[32] Ibid. pp.18-21.

[33] Ibid. p.21.

[34] Ibid. p.24.

[35] Ibid. p.25. A Recovery meeting may begin with an "endorsement" (giving oneself credit for one's efforts) followed by recitations by members of incidents that have an emotional upheaval, an enumeration of the symptom, displaying how the individual coped with the situation, and how Recovery helped. Part of the technique involved is in spotting each symptom as it arises, and especially "sabotaging" which takes the following forms: 1. literalness (blocking the physician's efforts by means of a literal misinterpretation); 2. ignoring or discrediting the initial improvement; 3. disparaging the physician; 4. challenging the physycian's diagnosis. Spotting is also used to spot emotionalism, sentimentalism and lack of muscle control. Examples of sentimentalism might be expressed in the following words: "I like to think of myself as a lost soul. I feel I am irretrievable". Emotionalism is demonstrated by a patient as follows: "I am disgusted with myself. I can't control my temper. The other day it was so bad that I cried out 'I am sick and tired of living'.

Of course, they were alarmed. I knew they would be."

[36] Ibid. p.25.

[37] Ibid. p.27.

[38] Ibid. p.28.

[39] Ibid. p.28.

[40] See Robert Kenneth Jones, "The Catholic Apostolic Church: a Study in Diffused Commitment", in Michael Hill, (ed.), A Sociological Yearbook of Religion in Britain, 5, London: SCM, 1972, pp.137-160, especially p.143: R.K. Jones, op.cit. (1970) pp.184-5; Robert Kenneth Jones, "The Swedenborgians, a Case of New Wine in Old Bottles; an Interactionist Analysis", in Michael Hill (ed.), A Sociological Yearbook of Religion in Britain, 7, London: SCM, 1974. See also Chapter Four, Part One.

[41] Henry Wechsler, "The Self-Help Organisation in in the Mental Health Field: Recovery, Inc., a Case Study", Journal of Nervous and Mental Disease, 130, No.4, 1960, pp.297-314.

[42] Ibid. p.302.

[43] Ibid. p.301.

[44] Ibid. p.307.

[45] Ibid. p.307.

[46] Charles Y. Glock and Rodney Stark, Religion and Society in Tension, Chicago: Rand McNally, 1965.

[47] David Silverman, The Theory of Organisations, London: Heinemann, 1970, pp.8-14.

[48] Ibid. p.9.

[49] Ibid. p.10.

[50] Tom Burns, "The Comparative Study of Organisations", in Victor Vroom, (ed.), Methods of Organisational Research, Pittsburg: University of Pittsburgh Press, 1967, pp.130-1.

[51] Although affiliated to the central organisation many branches of Neurotics Nomine prefer to call themselves by another name such as New Horizons Club, the Endeavour Club, and so on. Other organisations with a common interest are the Henry VI Society (with the aim of studying the life of the

king and praying for the re-opening of his cause in Rome: entry not restricted to Roman Catholics) and the Open Door principally for the estimated one hundred thousand agrophobia sufferers in Britain) and the Red Bridge of Friendship (in Australia). On a national scale there exists a correspondence club in which members can write to each other about their problems.

[52] Recovery, Inc. has developed differently from Neurotics Nomine in a number of ways; more explicit leadership; intense commitment; more homogeneous recruits; more procedures for instituting social control; more rigid and formal structure; liturgical character of meetings; charismatic element in the founder; structure tighter and more strictly defined; structured system of goals, methods and taboos; more bureaucratic structure, and so on. No doubt some of these differences are to be accounted for in terms of social and historical circumstances in development. Because of these characteristics the suggestion appears to be that Recovery, Inc. approximates more than Neurotics Nomine to the social structure and organisation of transcendental secular groups. If this is the case then we may be led to the implication that the social structure and organisation of groups, whether secular or transcendental, if the resemblance or similarity is sufficient, will be the over-riding factors in drawing out similarities, irrespective of the extent of the dissimilarities of the ideologies, goals, and so on.

[53] For an account of the therapeutics of fellowship, which these two groups share with Alcoholics Anonymous, see R.K. Jones, op.cit. (1970) pp.185-6. Also Chapter Three.

[54] Erving Goffman, Asylums, New York: Doubleday and Company, 1961, pp.62-3.

[55] B.R. Wilson, Sects and Society, 1961, London: Heinemann. See also Bryan Wilson, Religious Sects, 1970, London: Weiderfield and Nicholson, pp.2-35, for an up-to-date analysis regarding the problem of definition. The main characteristics of sectarianism appear to be voluntariness, exclusivity, merit, self-identifiction, expulsion, conscience, and legitimation. I am following here the yardstick which I adopted in an earlier study (R.K. Jones, op.cit. 1970) as follows: See Chapter Three.

"There is now quite general agreement concerning the characteristics of religious sects. Wilson (1961) defines sect as 'that small religious group in which membership is voluntary and conditional upon some mark of merit - understanding the group's teachings, or experience of some personal religious ecstasy - upon the basis of which association can arise'. The sect is small and exclusive, and its voluntary membership is generally adult...separation from society...exclusiveness both in attitude and in social structure (or characteristics)". Other characteristics include symbolism, voluntary joining, spirit of regeneration, single mindedness and fervour of the adherents, members kept under scrutiny and the pattern of their lives regulated in particular ways, equal total commitments and subjective fellowship and community.

[56] Partly because of the high level of commitment but also because such ideologies approximate to a total explanation, usually by applying one overriding idea to "the various realms of reality". For example, Recovery, Inc. bases its ideology on the principle that mental health can be both attained and sustained in day-to-day social interaction in the normal world by the systematic application of techniques relating to the mastery of the "will".

9 Conclusions

"...the special kind of behaviour that we call religious is composed of the same elements as is behaviour which is not religious." (MacMurray)

The starting point for this work lay in the similarities which were observed between groups organised around sets of activities but which were clearly claiming to be differentiated on the transcendental/secular continuum. A survey of the literature revealed that while many had commented on these similarities little had been accomplished in the way of a sustained study of such groups. Such studies that had been made appeared content to draw broad parallels rather than produce sustained and rather more penetrating analyses. The task was even more onerous because such groups as were chosen differed widely in their origins, growth and intentions, even though, as we contend, they resemble each other in many respects. Nevertheless, each group emerged in its own specific milieu and deeply embedded in particular historical circumstances.

This has been in many ways an exploratory study of a selected number of groups chosen partly because no sociological study has been done of most of them but also because they were all groups responsible for managing the identity of their members in some way and utilising some sort of ideological legitimation. It is true that other groups exhibiting both a secular and a transcendental "value-orientation" could equally have been examined - groups such as clubs for the Hard of Hearing, Weight Watchers, Gamblers Anonymous, innumerable political segments, the Aetherius Society, Colostomy clubs, groups for what Goffman terms the visibly handicapped, in fact a hundred and one other groups.

The structure of the groups which are secular or transcendental presents quite a number of problems simply because there appears to be little agreement on the meaning of the term "structure". Certainly "structure" involves such ideas as "fixed relationships" between the elements of the whole which relate to each other in particular ways. Some of the ways in which the elements of the social structure of groups can relate to one another are as follows:

1. Each social structure involves the assumption that a cognitive map such as beliefs and knowledge are accumulated and sustained within it.

2. There exists what Parsonians might term tension management processes for the communication of sentiment or feeling.

3. Social norms.

4. The allocation of status and role.

5. Activity, latent or manifest, directed towards the attainment of goals.

6. The existence of authoritative and non-authoritative power generally revealed in processes involving decision-making.

7. Positive and negative sanctions.

8. Stress-strain which may result in disorganisation.[1]

Furthermore, within social structures we have such fundamental processes as communication between actors, socialisation into the group, boundary maintenance which any particular social structure utilises to maintain its identity, process of social control for the regulation of behaviour, the social change which occurs in the social structure of groups over time, and institutionalisation whereby certain forms of behaviour become socially legitimated. Obviously, social groups resemble social systems because they are concrete manifestations of the latter and in modern sociology the term social structure is applicable not only to large associations but also to smaller groups.

However, some of the problems associated with terms such as "social structure" we have highlighted in Chapter One. Boudon[2] puts the point (and the dilemma) nicely when he says that "On one hand, there is a universal recognition...of the fundamental importance of this concept for the human sciences. On the other, (social scientists) cannot fail to recognise the ambiguity surrounding it."

Having indicated some of the complexities of the term "social structure" we then turned to our main consideration, of trying to account for apparent similarities of organisation and structure in secular and transcendental value-orientated settings. The association of "social structure" and "organisation" is again, unfortunately, fraught with difficulties. Some social scientists would clearly argue for an identification of the two terms whereas others see two quite distinct meanings. [3] Thus in one sense the groups we have examined have all the characteristics of overt organisation as well as a covert social structure. There are heuristic advantages in opposing the terms because it points us in the direction of behaviour other than the appearances.[4]

Several reasons were examined as to why secular, non-transcendental groups might resemble transcendental ones. Arguing from a review of secular parallels, most of which are too inclusivist to be of very much value as actual analyses of such groups, we suggested that many groups which are overtly secular both in their practices and their intentions deliberately appropriate "symbol forms" to facilitate the enactment of their own group programmes. The organisation and structure of certain ideological groups, irrespective of their stance being secular or transcendental, have in common the minimum organisational requirements for the facilitation of legitimating systems. Such legitimating systems are not only essential for any society but in advanced technological societies characterised by pluralism we have competing legitimating systems which might even result in a "privatisation" of belief, that is, "of the withdrawal of religious commitments from their traditionally designated locations in society..."[5] Hence a study of these other legitimating systems are necessary to arrive at a full picture of the phenomena which we term the sociology of religion. Such legitimating systems utilise a world view which is not in itself capable of direct observation of the behaviour of the individuals within the cosmological settings which house such world views.

The emergence of a common identifiable structure and organisation of groups such as ones we have discussed, which tacitly or implicitly manage the identities of their adherents, can be sketched out as follows. They generally begin by informal meetings of small groups of individuals who have in mind the same or similar objectives. In the case of A.A. this was Dr. Bob and Bill, with the Society of the New Church the small coffee meetings under the direction of Hindmarsh, with the SPGB the younger and more articulate members who led the revolt against Hyndman and the SDF, and with the Irvingites it was the Albury band under the direction of Drummond . Usually, in the initial stages, there is a lack of formal organisation, but the problems associated with promulgating the sect's teachings, with establishing a coherent ideology and a faithful core of followers, necessitates that rules be drawn and statuses distributed within the group itself. In this manner groups such as this very quickly develop a formal social structure and organisation and act as carriers of the ideology they put forward. Initiates have of necessity to be managed not only in ordinary face-to-face interaction but also in terms of their newly accepted identity. The former type of management includes what Garfinkel describes as the territories within which the interaction takes place, the renting of accommodation, elaborate timetables for meetings, agendas, rules for the way in which individuals must relate to each other, the exact nature of their tasks, the allocation of access, and so forth. The latter type of management is specifically concerned with the group's distinctive teaching and its internalisation. In this way the management of identity involves the perpetuation and inculcation of an ideology that is carried forward with a relative totality which permeates their social lives. Undoubtedly, there will be a degree of variance in the extent to which this pattern can be followed for while in our groups the early leaders in no sense were adopted as charismatic figures in other groups such as Buchmanism a certain element of personal charisma does exist. This is not to imply, of course, that a certain degree of reverence does not attach to the initial founders but in the case of our groups any real "devotion" was more likely to be transferred to the Writings such as those of Swedenborg, Marx, Alcoholics Anonymous, and the liturgical enshrinements of Irvingism.

Criteria for entrances and exits are speedily formulated during a period of structural differentiation. Deviants are expelled and recruits tested or asked to display some mark of merit in some way. Basically, structural differentiation is the way in which a group grows outwards whereby distinct segments adopt some specialised task. The kind of tasks such groups develop are fund-raising, recruitment, publishing, and so forth. Gradually, some of these enterprises may well be phased out, such as the recruitment policy of the Irvingites and to some extent the Swedenborgians and the SPGB. A.A. resembles much more a general service organisation in this respect because it never has recruited as such, merely offering itself as an available plausability structure for the reforming alcoholic.

The centralisation of organisation is noticeable in all our groups to a greater or lesser extent although some of them were relatively autonomous at group level. In most instances, not just of our groups but many others, the publishing enterprises are centralised on the assumption that whoever controls them also controls the ideology of the groups.

Commitment is not a uniform phenomenon but varies both within a movement or group and between movements and groups. Similarly, group ideologies mobilise individual and social discontent by offering a bridge between perceived problems and a general _Weltanschung_. Value-orientated belief systems involve an attempt at a reconstitution of a society to some degree. Such reconstitution involves some firmly established concept of _where_ the change is to take place and the _amount_ or degree of change sought. Thus some transformative value-orientated belief systems might aim at a _total_ change in the _social structure_, some at a _partial_ change, some at a _total_ change in the _individual_, and some at a _partial_ change in the _individual_. Our groups offer a sample of these categories because A.A. aims at a total change in the individual, the SPGB a total change in the social structure, while the Irvingites and the Swedenborgians aim at a partial change in both the individual and the structure.

An analysis of Alcoholics Anonymous revealed a secular movement which, in its organisation and structure, exhibited many of the characteristics of a transcendental ideology and shared a number of distinctly sectarian patterns. Similarly, a study of two overtly transcendental groups, the Society of the New Church and the Catholic Apostolics, gave a number of indications of how the identity of individuals was managed by settings which took particular stances towards the world and the interpretation of "reality".

Turning to a further secular value-orientated group, the Socialist Party of Great Britain, revealed that the "dynamics of sectarianism" were indeed to be found "at work in places far removed from religion proper". The SPGB's Party Ideology, despite the avowedly secular nature of the movement, as represented by the Principles does indeed display characteristics usually found in transcendental groups.

Although we have been arguing for an overlap between these two types of legitimating cosmological settings and for a number of similarities which might exist between them nevertheless there are important variations in terms of specific social processes that will serve to distinguish them. Such a distinction will not only mark off transcendental from non-transcendental but it will also serve to distinguish between the different patterns _within_ each category. Thus these groups will exhibit variations in terms of specific social processes such as origins, patterns of development or growth, degree of separation from the world, the extent of any activity which we might describe as proselytisation, and so on.

Such social processes are to be viewed not simply in terms of social interaction of members with each other but also with the external _milieu_. Because such groups exist in the first place this assumes that particular forms of social interaction occur[6] which constitute their dynamic aspects. Here again, with this term, there exists a great deal of vagueness. Generally, it appears to denote virtually any aspect of the social interaction process. Particularly, it seems to be used sometimes to indicate a period of transition, over time, from one social condition to another social condition. Social process is a term which suggests a macroscopic analysis of the dynamic aspects of interaction in social settings.

Variations within and between groups with respect to the social processes will include co-operation (formal, informal, and symbiotic), opposition (competition, conflict and rivalry), accommodation (domination, compromise, toleration and truce), assimilation, confrontation (formal, informal and adjustment), and homogenisation. Thus the groups we have looked at will vary in the degree that they co-operate with society and the members with one another. Such co-operation can be spelled out in terms of specific rights, duties and obligations (formal) or it can take place in an informal manner. The degree and manner to which social processes are actively taking place will determine the extent to which groups will differ from one another. Such co-operation can be spelled out in terms of specific rights, duties and obligations (formal) or it can take place in an informal manner. The degree and manner to which social processes are actively taking place will determine the extent to which groups will differ from one another.

In their origins our groups have exhibited distinct variations. Alcoholics Anonymous, Recovery, Inc., and Neurotics Nominee appear to be groups lacking in any charismatic leader. Edward Irving, however, appeared to be very much such a leader and similarly Swedenborg appears as a charismatic leader in absentio. On the other hand the SPGB seems devoid of any real charismatic elements in its origins. One interesting feature of the groups we have looked at is that none of them could be described as proselytising groups in the generally accepted sense of active engagement in seeking recruits. The Catholic Apostolics are quite the opposite, and are in fact isolated almost completely not only from other sects and from the broader issues of society but also their own remaining groups have little contact with each other. The Society of the New Church, apart from its half-hearted attempt at advertising has maintained a relatively contented position as a group willing to engage in "rational" discussion about its beliefs with others. Similarly, the SPGB has withdrawn from anything like the active stance which it displayed at one time in its development. Alcoholics Anonymous in no sense "goes out" and recruits members but is content to offer its services when individuals are designated as requiring them, and is consequently referred to by other agencies such as the medical profession.

The negotiation of personal meaning-systems within the groups we have examined is not only sustained by the structure of the groups but in many respects affects such a structure because of the hopes, values and conceptions entertained by their members. There is no inexorable logic implied in the structure of such groups because they *could* be other than they are, albeit within the bounds imposed by the minimum organisational requirements for the facilitating of legitimatory systems. Arising out of the vagueness of thinking of goals as something to be achieved by the organisation and structure of groups it might be more profitable to conceive of them as part of the process of legitimation which arises from the preoccupations and interests which the members of these groups entertain. The regularities and similarities exhibited by groups whose concern is principally to manage the identity of their members in some way can be explained by thinking of them as arising from the consequences of the negotiation for personal meaning systems which in turn can be firmly placed in the social interaction and intentions of each particular group. To some extent, also, such groups not only *become* a social reality for members but sustain such a construction by such devices as heavy emotional utilisation of salient symbols.

The scheme which we offered was a possible account of how our groups, value-orientated management groups, organised and structured their beliefs. Participation and affiliation to groups is usually classified under either "strain" or "interest" theories and recruits bring with them their specific "subjective" and "objective" constructions of reality. Such constructions of reality takes place because objective reality exists *in* the groups, settings and organisations which are external and pre-existent to the individual, and contrasted with this we have the subjective reality or the "internal logic" which actors ascribe to their confrontation with objective reality, and, of course, to the significant others in the latter. Both these constructions are synthesised into symbolic universes of meaning. Disregarding the goals-means dichotomy we were led to construct a taxonomy of theoretically polar opposites which are connected by a continuum. This, in effect, becomes the belief structure of the groups under discussion, or, to adopt a more dynamic and interactive approach, "territories of interaction".

This scheme is not, of course, the only one which could be constructed in trying to account for similarities in the organisation and structure of the six groups we have examined. For example, we could give a description of any one individual career in becoming affiliated to a group as follows:

The Acquisition of Group Structure

availability	inculcation of norms and attitudes
predisposition	internalisation of norms and attitudes
synchronic agreement in encounter	sustaining of acquired identity.

In other words the particular group structure was available to the individual who in turn, by being predisposed to actual joining, was himself available to the group. On encounter each would see the other as an agreeable agent in the encounter. Once contact is made the group must in some way inculcate its norms, values, attitudes, roles and so on to the individual who must internalise them. The remaining problem is to sustain the identity of the recruit as a member of the group which presumably supplies an adequate "plausability structure".

Another way of approaching the problem might be to ask: how are recruits to a secular or transcendental group socialised and how do they assume a sense of social structure? An ethnomethodological or symbolic interactionist perspective might view the process in terms of a microlinguistic analysis of the acquisition of group structure as follows:

A Possible Micro-Linguistic Analysis of Group Structure

reciprocity or interchangeability of perspectives	et cetera assumptions	retrospective-prospective sense of occurrence (interpretive procedures)
normal forms of acceptable talk	tacit commitment to normative order	display of intentions through commitment

This means that the recruit can interchange his standpoint with that of the group (members). His problems, whether of the overthrow of capitalism, psychological disturbances of personality, alcoholism, or deeply rooted feelings about the nature of the world will be understood by, and be intelligible to, other members because they share a common taken-for-granted world. Further, the _et cetera_ assumptions allow respective members to "fill in" vaguenesses and ambiguities, to overcome possible misunderstandings, and so on. Each member shares an expectation of what passes for normal talk and behaviour in these groups, and there will be a tacit commitment to the normative order. Conversations within these groups will rely somewhat on interpretive procedures, by which is meant that the social structure of the group will be sustained by individuals "reading in" meaning and understanding into terms that may not be immediately apparent. This may well be a process of "Waiting for later utterances (which may never come) to clarify present descriptive accounts, or 'discovering' that earlier remarks or incidents now clarify a present utterance..."[7] Finally, a member or recruit to groups with specific ideologies and programmes declares his future intentions through his act of commitment to that group.

This study has attempted to pinpoint some of the reasons why groups, overtly dissimilar in scope and doctrinal content, do in fact resemble each other in a number of ways. The groups we chose were categorised as managing or controlling the identity of their members in some way which in itself imposed a number of constraints on the form that the structure and organisation would take. Our point was further reinforced by the conclusion that irrespective of the ideological stance being secular or transcendental such groups held in common the minimum organisational requirements for actually being legitimating systems.

NOTES TO CHAPTER NINE

[1] Alvin L. Bertrand, "The Stress-Strain Element of Social Systems: a Micro-Theory of Conflict and Change", Social Forces, Vol.42, No.1, 1963. Also, Charles P. Loomis, Social Systems: Essays on their Persistence and Change. 1960, New York. Van Nostrand.

[2] Raymond Boudon, The Uses of Structuralism. 1971. London. Heinemann, p.1.

[3] For example, Kroeber has argued for an identification of "organisation" and "social structure" whereas Gurvitch has indicated the usefulness of the distinction. See A.L. Kroeber, Anthropology New York, 1948, and Georges Gurvitch, "Le Concept de Structure Sociale", Cahiers Internationaux de Sociologies, 19, 1955, pp.3-44.

[4] Boudon, ibid. p.42.

[5] Peter Berger and Thomas Luckmann, "Sociology of Religion and Sociology of Knowledge", Sociology and Social Research, XLVII, No.4, July 1963.

[6] Robert E. Park and Ernest W. Burgess, Introduction to the Science of Sociology, 1951, Chicago. University of Chicago Press.

[7] This approach is to be found in the various works of Alfred J. Schutz, and particularly Aaron V. Cicourel, "The Acquisition of Social Structure" in Jack D. Douglas Understanding Everyday Life, London. Routledge and Kegan Paul, 1971.

Appendix A: Transsexualism

1st Drunk: "Who was that woman I saw you with last night?"
2nd Drunk: "That was no woman; that was my husband!"

Mol (1976, pp.142-165) devotes an entire Chapter to a discussion of the "Self and Sexuality", referring to sexuality as the "actual foci of identity", and as a constituent of the "primary foci" of identity. Thus he writes: "It is not accidental that sex is now widely regarded as a primary source of personality integration." (p.155) He quotes Eliade's (1967, p.175) statement on the androgynous and hermaphroditic myths where androgyne is related to a "general formula signifying autonomy, strength, wholeness; to say of a divinity that it is androgyne is as much as to say that it is the ultimate being, the ultimate reality".[1] This is a position far removed from that of St. John Chrysostom, a Greek Christian Father and Patriarch of Constantinople in the second half of the fourth century A.D., who demonstrated his resiliance to sexual temptation from the opposite sex by dispatching a woman off a cliff. Sexuality highlights very clearly, according to Mol (p.155) the competitive nature of the relationships between personal, group and social identity. The sacralisation of a collective or shared identity places an emphasis on a restrained sexuality whereas an emphasis on individual identity "stresses the sacredness of sexual fulfilment".(p.156)

The synthesis of identity theory and development reaches a substantive area in the literature on gender identity and more particularly in a discussion of transsexualism related to psychotherapeutic techniques. This essay will explore some of the mechanisms of gender identity and gender reassignment using Mol's identity theory of religion to illuminate the processes at work, namely, those elements of sacralisation which are termed commitment, objectification, ritual and myth. Transsexualism is chosen as a fertile area of study because it brings together the subjective and objective aspects of identity theory as well as offering a nucleus of social and personal identity. It is in this area, too, that the biological and the sociological interact and complement each other.

Gender identity, in psychiatric and psychoanalytic terms of reference, becomes established within the first five years of life (Lidz, 1968, p.301). It is an integral part of adult identity that individuals clarify and strengthen gender identity. The substantive identity factor is the gender determinant of male or female, boy or girl. It is perhaps the "single most important determinant of personality characteristics. The difference is more than a matter of anatomy. The two sexes have different ways of relating and reacting, different perspectives and interests, different tasks and roles. The human species is divided into two sexes, and in every culture the males and females have different tasks and functions though the division varies from place to place" (Lidz, p.208). Lidz quotes approprietely from Freud (1933): "Male and female is the first differentiation that you make when you meet another human being, and you are used to making that distinction with absolute certainty." Psychoanalysts would argue that a firm identity with one's own sex membership is essential to both emotional stability and the development of a cohesive ego identity. The biological basis of sexual identity establishes the gender identity although subsequent "displacement" in terms of various sexual objects to enable sexual gratification can be encouraged by a range of interpersonal factors. Subsequent behaviour and orientation as "male" or "female" is considered by some writers such as Erikson (1959) to be related to the biological structure of the sex organs themselves, so that a boy develops an active life stance whereas a girl "develops feelings and ideas of an inner space". In later life some, such as Melanie Klein, maintain that early genital experiences can affect adult relationships. More often than not, however, "the assigned sex concurs with chromosomal and anatomical sex except under unusual circumstances" (Lidz, p.209). The poet Rilke, for example, was brought up as a girl by his parents throughout his early childhood.

The biology of sex gender

The biological determinants of sex gender lie in the presence of an X or Y chromosome in the fertilising agent and the joining of this with the X chromosome in the ovum.

> The presence of the Y chromosome usually assures that at a critical phase in fetal development the testes secrete androgen which influences the anlage of the genito-urinary tract to develop male internal organs and later to develop male external genitalia...the hormone acts upon the undifferentiated brain to organise certain circuits

in male rather than female patterns...It seems unlikely that androgen secretion in the mother would induce such patterned male behaviour in human infants...men with Klinefelter's syndrome, an anomaly due to the presence of two X chromosomes and one Y chromosome...may tend towards confusions in gender identity."
(Lidz, pp.210-211)

Evidence points to a predisposition at birth to a male or female gender orientation but that such a predispositional pattern can be modified by subsequent socialisation procedures.[1] Subsequent changes in gender assignment after the age of two and half "are very likely to create serious problems for the child" not least because by that age "the identity of the child as a boy or girl is already well ingrained in the child's awareness and behaviour" (Lidz, p.212). The child thus develops a sex identity initially through a process of modelling where he uses his parents as identification figures. The establishment of sex-role standards by assimilation from the culture in which he is raised involves a series of role-playing activities as part of the total process of sex-role identification. Eventually the individual regards or perceives himself as feminine or masculine which is the stage of sex-role identity.[2]

The transsexual

The term "transsexual" refers to an individual's identity as male or female not corresponding to their anatomical sex (Cauldwell, 1947; Benjamin, 1953). Transsexuals feel distinctly ill at ease living in the sex role which corresponds to their anatomy (Oles, 1977, p.67). The difference between homosexuality and transsexualism is simply that "The criterion of homosexuality...is sexual behaviour involving individuals of the same sex, while the criterion of 'inversion' is a personality in which the person's thinking, feeling and acting are typical of the opposite sex" (Brown, 1958, p.424). Gender identity forms a continuum of sex and gender role orientation formulated by Benjamin (1966) as a seven category Sex Orientation Scale.[3] The term gender dysphoria syndrome (Fiske, 1973) describes the presentation of the symptoms and biographies of those desiring sex-reassignment. The physical and psycholosocial aspects of sex-reassignment "usually include hormone treatment with masculinising or feminising hormones, changes in clothing and hair style, voice practice with a tape recorder and speech therapist, electrolysis for beard removal in male-to-female transsexuals, and, finally, surgery" (Oles, 1977, p.69). The stages in transformation outlined by Oles reveal the order of identity change. "When a person begins taking hormones he or she is beginning to make the sex change a reality and sometimes finds that the reality does not match the dream...There is the realisation that the sex change will not be a total transformation of either personality or physique" (p.69). When finally a decision is made to "declare" one's adopted identity then "The decision to cross-dress in public is an important step in the sex change process...By appearing in the clothing of the desired sex-role the transsexual is publicly expressing his or her identity and experiencing other people's reactions to it" (p.69).

Conversion surgery

Surgery becomes the final and irrevocable step in sex reassignment although such "conversion surgery" is only a "confirmation and completion of what has already proved to be successful" (Money and Gaskin, 1971)[4]. The presentation of the self for surgery invariably requires the fulfilment of three criteria: a) the desire to dress in the manner of the opposite sex; b) seeking to be regarded and socially accepted as a member of the opposite sex; c) and a persistent desire to undergo a "conversion operation" (Lukianowics, 1959). The prevalence of transsexualism in the population is difficult to guage.[5] Before gender reassignment was surgically performed on a group of Singapore transsexuals they had to demonstrate that they had never had any heterosexual inclinations, that they had been living the life of a woman for the last three years, and that they were psychologically female in every respect (Tsoi, 1977). Of the 56 cases in the Singapore sample 58% underwent "the sexual reassignment operation, which consisted of at least penectomy, castration and construction of artificial vagina and female external genetalia. Some cases were followed by operations to enlarge their breast"(p.408). Certainly for the Singapore transsexuals their condition "did not appear to be related a) to the lack of father figure or undue attachment to the mother, as reported by Stoller (1968) or, b) to cross-gender assignment and rearing" (p.408). Socarides (1969) appears to argue in an opposite fashion that "there is no evidence that gender identity confusion - a gender identity contrary to the anatomical structure - is inborn"(p.131). In fact, and perhaps not surprisingly, those who advocate the inborn nature of transsexuality also tend to be supporters of surgical conversion, whereas those who assign its aetiology to sociological variables tend to be advocates of psychotherapeutic treatment. The aetiology, it is argued, is not treated by surgical and endoctrinal methods. "Are those patients free of their former conflicts, no longer homosexual, transvestitic, or schizophrenic, or do they now pursue these drives without conflict, having only altered their external anatomy?"(p.131).[6]

Transsexual biographies

The biographies of transsexuals such as Agnes (Garfinkel, 1968), Jane Fry (Bogden, 1974), Roberts Cowell (Cowell, 1954), Jan Morris (Morris, 1974), and Christine Jorgensen (Jorgensen, 1967) are possibly more illuminating than the clinical data of a number of psychiatric studies. Agnes, for example, was absolutely convinced that she was a woman born with a man's body. "Much of the instrumental realism that she directed to the management of her chosen sexual status was concerned with so managing her circumstances as to avoid what she treated as a mistaken and degrading identity" (Garfinkel, p.130). An elaborate system of passing was developed after Agnes's commitment to an external signification in the form of feminine attire at the age of seventeen. Jane Fry voices similar sentiments when she says, "I am a female with a birth defect. I am a woman, but I have the organs of the other sex. I want to be a whole person again" (Bogden, 1974, p.21). Jane asks, "What is a person? Is a person what he is on the inside? Or what he is on the outside?...The only thing I want to change is my body, so that it matches what I am"(p.19). Roberta Cowell writes that "For the first thirty years of my life I was Robert Cowell, an aggressive male who had piloted a spitfire during the war, designed and driven racing cars, married and become the father of two children. Since May 18th, 1951, I have been Roberta Cowell, female. I have become a woman physically, psychologically, glandularly and legally" (Cowell, 1954). After a commitment to his feminine gender identity and to a surgical operation Cowell had to begin a long and difficult period of rehearsal in which he practiced behaving like the woman he so much wanted to be. Sometimes, though rarely, in the area of sex hormone interaction on the pathways of the brain, "a conventionally masculine or feminine person suddenly switches completely over to the other sex role" as in cases of extreme senility or during a rare type of temporal lobe epileptic seizure" (Money and Tucker, 1976, p.74).

Gender identity, then, is an integral aspect of adult identity, of the sense of oneself as male or female. It is "The sameness, unity and persistence of one's individuality as male, female or ambivalent, in greater or lesser degree, especially as it is experienced in self-awareness and behaviour" (Money and Ehrhardt, 1972). It is a reflective inner experience, private to the experiencing individual and generally unknowable to the observer. Gender role "includes everything you feel and think, everything you do and say, that indicates - to yourself as well as to others - that you are male or female. Gender identity and gender role are not two different things; they are different aspects of the same thing..." (p.9). The newborn are wired but not programmed for gender identity and the gender identity option is open at birth even for normal infants (p.91) although after the age of three or four "When the gender identity gate closed behind you it locked tight"(p.119).

To have one's gender identity reassigned initially by oneself and eventually by others in the form of a public legitimation is the ultimate transformation of identity. One might argue that death is the ultimate transformation of identity and, paradoxically, the ultimate denial and extinction of identity. However, as there is no enduring "I" to produce a reflective commentary on the transformation from life to death and the dialectic quality of both a personal and social identity cease then it is more profitable to regard death as the denial or negation, rather than the transformation, of identity. Only in the cases of mummification do we come close to a posthumous preservation of identity, in this instance by the action of the bereaved to maintain the identity of the deceased (Gardner and Pritchard, 1977).

Mechanism of sacralisation

Conversion surgery becomes the final step in a long process of identity change. Gender reassignment is, however, an event which can be usefully examined in the light of Mol's account of the four mechanisms of sacralisation which will enable us to describe it finally as a process of alternation which utilises the mechanisms of sacralisation to come to "terms" with the ultimate question of "Who am I?" Simply stated, transsexuals feel at variance with their sex role which corresponds to their physical anatomy. Such variance often amounts to disgust and sometimes to self-mutilation. The outward manifestation of sexuality is incongruous with their gender identity, and consequently "The goal of the transsexual is usually to undergo a physical transformation into the gender of choice, and to live and be accepted in the desired gender role" (Oles, 1977, p.67). The developmental period of sexual transition is one of extreme sensitivity to external reactions, and transsexuals "may search for role models, sometimes idealised ones, among those in the desired gender" (Oles, p.69). The fear of being "read" (recognised as transsexual) is paralleled by "gays" pretending to be "straight" in a seemingly never-ending world of secrecy which constitutes the homosexual community. The declaration to "come out" or "announce" themselves renders the stigmatised liable to discreditability. The transsexual identity, during the period of physical transformation, is protected by sacralisation which is "the process of safeguarding and reinforcing a complex of orderly interpretations of reality, rules and legitimations" (Mol, 1976, p.202).

The first of the four mechanisms, objectification, is the tendency to "sum up variegated elements of mundane existence in a transcendental frame of reference" (Mol, 1976, p.206). Objectification occurs for the transsexual when he pieces together a number of elements from a precarious and indeterminate existence in order to construct an external jigsaw puzzle that will also give him internal validity as a "whole" person. The transcendental and objectified goal becomes the desire for a complete sexual transformation. As Parsons (1968, p.14) says: "First, to function adequately in the psychological and sociological contexts and in their continual interplay with each other, the personality of the individual must be definable as a clearly established and focused <u>object</u>, both to the individual himself as answers to the questions of who and what he is and to the others with whom he interacts." Such a process is an integral part of <u>any</u> human cultural-level social interaction where the individual is both <u>actor</u> and <u>object of orientation</u>. Thus years of questionings and doubt find an external and objective framework.

The second of Mol's mechanisms of sacralisation is that of commitment which he defines as "focused emotion or emotional attachment to a specific focus of identity. It is an anchoring of the emotions in a salient system of meaning..."(p.216). Commitment can be secular or transcendental, with a "variety of identity foci" but there is invariably "a close relationship between commitment and consistency of behaviour"(p.217). The goal of the transsexual is gender conversion surgery. Following from the conviction that he or she is "basically" a member of the opposite sex there stems an overwhelming commitment to a number of means of obtaining that goal. Such commitment is intense and all-consuming.

The mythical, or third element, contributes "to the resolution of basic conflicts through objectification" (p.246) though as such is independent of both ritual and commitment. Perhaps the key lies in the statement that "myths provide the fitting contour for existence, and so hold arbitrariness and meaninglessness at bay" (p.253), and "disconnected experiences are reordered into an emotionally anchored unit"(p.261). The transsexual entertains many myths not least of which is the belief that conversion surgery will solve all his problems. Some, however, such as Socarides (1969) have argued that the intervention of surgery merely sanctions the transsexual's pathological view of reality and that it cannot resolve the underlying conflict. Gay mythology and ideology (Warren, 1974) have their counterpart in the transsexual universe, where a number of unrealistic idealised role models are held, for example, that the bi-sexual is the pinnacle of creation. Oles has reported some mythology as follows (1977, p.72): "Some male-to-female transsexuals have expressed the belief that women lead irresponsible, dependent lives, and that once they become women there will be men waiting to take care of them. Others have made statements such as, 'I know I'm a woman because men aren't sensitive and prone to cry as I am.' Some individuals equate femininity with flawless beauty and movie-star-like appearance, and become depressed when they cannot reach this ideal". Garfinkel's Agnes (p.128) claims "I have always been a girl", which is accompanied by a "remarkably idealised biography in which evidences of her original femininity were exaggerated". Striving towards what she would like to become Agnes "had an abiding practical preoccupation with competent female sexuality"(p.121). She envisaged a real world of "sexed persons" as a matter of objective, institutionalised fact.[7]

The fourth mechanism is that of ritual. Rites perform a number of functions. They "articulate and reiterate a system of meaning, and prevent it being lost from sight. They act out a sacralized sameness. They restore, reinforce, or redirect identity" (Mol, 1976, p.233). Identity changes take place in extremely ritualistic contexts. With the transsexual public cross-dressing is the culmination of a thousand private rehearsals, an expectation of social validation from the "normal" world. The conviction that one is a transsexual is accompanied by concomitant behaviour deemed appropriate to the opposite sex, e.g., "alteration in dress, interests, attitudes, choice of sexual object, etc." (Socarides, 1969, p.125). It is, in fact, "The choice of clothing and hair style and the adoption of certain mannerisms (which) reflect the type of person the transsexual feels he or she is, or wants to be, in the gender of choice" (Oles, 1977, p.70). Agnes was more than susceptible to the ritualistic elements of femininity, for example, "holding her arm while I guided her across the street...offering to hang up her coat; relieving her of her handbag; holding the automobile door for her while she entered...It was on such occasions that she most clearly displayed the characteristics of the '120 per cent female'. At such times she acted like a recent and enthusiastic initiate into the sorority of her heart's desire" (Garfinkel, 1967, p.133). The stereotypes which are caricatured relate to an idealised desired sex rather than an actual one. Some transsexuals emphasise the reinforcing qualities of ritualised gender responses. "The very tone of voice in which I am now addressed, the very posture of the person next in the queue, the very feel in the air when I entered a room or sat at a restaurant table, constantly emphasised my change of status...The more I was treated like a woman, the more woman I became..." (Morris, 1974, p.120, in Money and Tucker, 1976, p.123). Consequently, they sound "not like a woman, but like a man's idea of a woman...like a woman in a TV commercial" (West, 1974, in Money and Tucker, p.206).

I have attempted to apply Mol's schema of the mechanisms of sacralisation, of identity maintenance or what Parsons would designate as "the pattern-maintenance code-system", to the problems of gender reassignment. It is, of course, debatable whether a macrotheory of religion can have elements capable of being related to a problem which is essentially genetic-psychotherapeutic. Nevertheless, a process of emotional welding is taking place in the transsexual context without parallel in intensity and heart searching, and indeed without parallel in terms of purely physical conversion. The change in personal identity also involves a change in the social identity. How the transsexual views himself will be translated into action and appearance and will be experienced by others. Secular sacralisation processes occur all around us; in the alcoholic attaining sobriety, in the drug addict "kicking" the habit, in the psychotherapeutic "switch", and so on. The loosening of previous allegiances (Strauss, 1959, p.123) and the concomitant search for secular identity results in a building up of a new identity which I have argued takes place along the lines of Mol's schema. The seeming immutability of one's own sex when confronted with the knowledge that transsexuals _do_ exist brings into question the very _meaning_ of what it is like to be a male or female. The act of changing one's sexual identity, of the reassignment of identity, is a necessary area in any symposium on identity theory.

NOTES TO APPENDIX "A"

[1] The studies by Hampson and Hampson (1961) and Money (1957; 1965) cited in Lidz (1968, pp.211-212) deal with a sample of seventy-six pseudohermaphrodites and hermaphrodites. "The cases studied include females with virilizing adrenal cortical hyperplasia, simulant females with testes, chromosomal males with gonadal agenesis and female physical morphology, cryptorchid males reared as females, and others. There were contradictions between chromosomes and the gender in which the child was reared in nineteen cases; between the gonads and the gender...in twenty; between hormonal sex and the gender in which the child was reared in twenty-seven postpubertal cases; between appearance of external genitalia and the gender in which the child was reared in twenty-three; and between internal accessory organs and gender rearing in eighteen".

[2] A number of problems present themselves which cannot be fully discussed here, although we can in fact raise them. Some writers, for example, do not differentiate between non-transsexual and transsexual homosexual males and conceive of the former as a disorder comparable to an overvalued idea. In one study all fifty-two applicants for sex reassignment surgery stressed a preference for heterosexual rather than androphilic male partners. Feminine gender identity, if used as a criterion for reassignment, is also difficult to estimate and in homosexual males transsexualism appears to be an extreme of a frequently occurring attribute (Loraine, 1974, pp.54-55). High feminine gender identity is not confined to some homosexual males but is also present in some heterosexual ones. In biological studies "It appears that at least under natural conditions copulatory interaction between sub-human primate males serves mostly the function of signalling non-sexual agnostic relationships" (Loraine, 1974, p.62). The interplay of constituents such as the chromosomes, hormones, reproductive glands and upbringing are the determining factors in gender identity.

[3] This relates to cross-dressing and ranges from: unpleasant, pseudo-transvestite, fetishistic transvestite, true transvestite, non-surgical transsexual, true transsexual of moderate intensity and of high intensity (Benjamin, 1966).

[4] For an interesting summary of physicians' attitudes to transsexual conversion surgery see Baker (1969). Merloo (1967) states that "the project of sex reas-

signment has been launched with all the paraphernalia and fanfare of scientism. There is no evidence whatsoever that adult transsexualism can be 'cured' by behavioural techniques although there appears to be some success with childhood transsexualism before the age of nine."

[5] Tsoi (1977) estimates the demand for transsexual male reassignments as 1 in 25,000. Pauly (1968) in America estimated the demand as 1 in 100,000, whereas Walinder (1968) estimated the figure to be 1 in 37,000 in Sweden, while Hoenig and Kenna (1974) put the figure as 1 in 34,000 for England and Wales. Hamburger's (1953) research put the figure as 3 to 1 for male to female reassignments. Tsoi's Singapore Transsexuals appeared to reach their inception around 22.0 years of age, were mostly engaged upon feminine occupations, were above average in educational attainment, nearly half (43%) were married and 24% had fathered children. In psychological testing the mean score was 105 and in the MMPI was nearly one standard deviation above the British normal population (Tsoi, 1977).

[6] Stoller (1967; 1968) has elicited the following factors as sociologically conducive to transsexualism: excessive contact and gratification by the mother; the mother's bi-sexuality; low rate of divorce despite empty marriage bed; dynamically absent father; and a high order of unusual permissiveness.

[7] Socarides (1969), quoting Worden and Marsh (1955) says that "these individuals had no conception of the duties or responsibilities entailed by being a woman but were, rather, wrapped in fantasies of being beautifully dressed, embellished with sparkling jewelry, wonderful coiffures, cosmetics, etc".

****

Benjamin, H., "Clinical Aspects of Transsexualism in the Male and Female", American Journal of Psychotherapy, 18, 1964, pp.458-469.

Benjamin, H., The Transsexual Phenomenon, New York: Julian Press, 1966.

Benjamin, H., "Should Surgery be Performed on Transsexuals?", American Journal of Psychotherapy, 25, 1971, pp.74-82.

Bogden, R., (ed.), Being Different: The Autobiography of Jane Fry, New York: Wiley, 1974.

Brown, D.G., "Inversion and Homosexuality", American

Journal of Orthopsychiatry, Vol.23, 1958, pp.424-429.

Cauldwell, D., "Psychopathia Transexualis", Sexology, 16, pp.274-280.

Cowell, R.E., Roberta Cowell's Story, London: Heinemann, 1954.

Eliade, M., Myths, Dreams and Mysteries, New York: Harper, 1967.

Erikson, E.H., "Growth and Crises of the Healthy Personality", Psychological Issues, Vol.1, No.1, Monograph No. 1 New York: International Universities Press, 1959.

Fiske, N., "Gender Dysphoria Syndrome (the how, what and why of a Disease)" in Proceedings of the Second Interdisciplinary Symposium on Gender Dysphoria Syndrome, (eds. Laub, D., and Gandy, P.) Stanford, California: Stanford University Press, 1973.

Freud, S., "The Psychology of Women", New Introductory Lectures on Psychoanalysis, New York: W.W. Norton, 1933.

Gardner, A. and Pritchard, M., "Mourning, Mummification and Living with the Dead", British Journal of Psychiatry, Vol.130, 1977, pp.23-28.

Garfinkel, H., Studies in Ethnomethodology, Englewood Cliffs, New Jersey: Prentice Hall, Inc., 1967.

Hamburger, C., "The Desire for Change of Sex as Shown by Personal Letters from 465 Men and Women", Acta Endocrinologica, 14, 1953, pp.361-375.

Hampson, J.L. and J.G., "The Ontogenesis of Sexual Behaviour in Man", in Young, W.C. (ed.) Sex and Internal Secretions, Vol.2, 3rd edition, Baltimore: Williams and Wilkins, 1961.

Hoenig, J. and Kenna, J.S., "Prevalence of Transsexualism in England and Wales", British Journal of Psychiatry, 124, pp.181-190.

Jorgen, C., Christine Jorgensen: a Personal Autobiography, New York: Paul S. Erickson, 1967.

Lidz, T., The Person, New York: Bsic Books, Inc., 1968.

Loraine, J.A., ed. Understanding Homosexuality:

its Biological and Psychological Bases, Lancaster: MIP, 1974.

Lukianowicz, N., "Survey of Various Aspects of Transvestism", Journal of Nervous and Mental Disease 128, 1959, pp.36-64.

Merloo, J.A.M., "Change of Sex", American Journal of Psychiatry, 124, 1967, pp.203-204.

Mol, H., Identity and the Sacred, Oxford: Blackwell, 1976.

Money, J., Hampson, J.G. and Hampson, J.L., "Hermaphroditism: Recommendations Concerning Assignment of Sex, Change of Sex, and Psychologic Management", Bulletin Hopkins Hospital, 97, 1955, pp.284-300.

Money, J., Hampson, J.G., and Hampson, J.L. "Imprinting and the Establishment of Gender Roles", Archives of Neurology and Psychiatry, 77, 1957, pp.333-336.

Money, J. "Psychosexual Differentiation", in Money, J., ed., Sex Research, New Developments, New York: Holt, Rinehart and Winston, 1965.

Money, J. and Gaskin, R., "Sex Reassignment", International Journal of Psychiatry, 9, 1971, pp.249-269.

Money, J. "Gender Role, Gender Identity, Core Gender Identity: Usage and Differentiation of Terms", Journal of the American Academy of Psychoanalysis, 1, 1973, pp.397-403.

Money, J. and Ehrhardt, A., Man and Woman, Boy and Girl: the Differentiation and Simorphism of Gender Identity from Conception to Maturity, Baltimore: Johns Hopkins Press, 1972.

Money, J. and Tucker, P., Sexual Signatures, London: Harrap, 1976.

Morris, J. Conundrum, New York: Harcourt Brace Jovanovich, 1974.

Oles, M.N., "The Transsexual Client: a Discussion of Transsexualism and Issues in Psychotherapy", American Journal of Orthopsychiatry, 47, 1, 1977, pp.66-74.

Pauly, I.B., "Male Psychosexual Inversion: Transsexualism: a Review of 100 Cases", Archives of General Psychiatry, 1, 1965, pp.172-181.

Pauly, I.B., "The Current Status of the Sex Change Operation", Journal of Nervous and Mental Disease, 147, 1968, p.460.

Parsons, T., "The Position of Identity in the General Theory of Action", in Gordon, C., Gegen, K.J., eds. The Self in Social Interaction, New York: John Wiley, 1968, pp.11-23.

Socarides, C.W., "The Desire for Sexual Transformation: a Psychiatric Evaluation of Transsexualism", American Journal of Psychiatry, Vol.125, 1969, pp.125-131.

Stoller, R.J., "Etiological Factors in Male Transsexualism", Transactions of the New York Academy of Science, 29, 1967, pp.431-432.

Stoller, R.J., "Parental Effects in Male Transsexualism", in Green, R., ed. Transsexualism and Sex Reassignment, Baltimore: John Hopkins University press, 1968.

Stoller, R.J., Sex and Gender: On the Development of Masculinity and Feminity, New York: Science House, 1968.

Strauss, A.L., Mirrors and Masks, the Search for Identity, Glencoe, Illinois: Free Press, 1959.

Tsoi, W.F., Kok, L.P., and Long, F.Y., "Male Transsexualism in Singapore: a Description of 56 Cases", British Journal of Psychiatry, Vol.131, 1977, pp.405-409.

Walinder, J., "Transsexualism, Definition, Prevalence and Sex Distribution", Acts Psychiatrica Scandinavica, Supplement 203, 1968, pp.255-258.

Warren, C.A.B., Identity and Community in the Gay World, New York: John Wiley and Sons, 1974.

West, R., Book Review of Morris (above), New York Times Book Review, April 14, 1974.

Worden, F.G., and Marsh, J.T., "Psychological Factors in Men Seeking Sex Transformation: a Preliminary Report", Journal of the American Medical Association, 157, 1955, pp.1292-1298.

Also:

Baker, H.J., "Transsexualism - Problems in Treatment", American Journal of Psychiatry, Vol.125, 1969, pp.118-124.

Benjamin, H., "Transvestism and Transsexualism", International Journal of Sexology, 7, 1953, pp.12-14

Appendix B: Death and identity

Introduction

Although death comes to us all it manifests itself in different disguises and appears along different pathways. Such differentiation is historical and sociological for the simple reason that:

> The social body constrains the way the physical body is perceived. The physical experience of the body, always modified by the social categories through which it is known, sustains a particular view of society...The care that is given to it... must correlate closely with the categories in which society is seen.
> (Douglas, 1973: 93)

Kleinman (1973: 206-213) takes this idea further when he describes medicine as having a "symbolic reality" whose "systems function along the lines of the cultural dialectic, relating and treating both individual and social realities...A given medical system...structures the experience of illness and, in part, creates the form disease takes". We will argue that death, also, is a social construction, which in turn affects our style of dying and the values and meanings which accrue to it. Death challenges the "taken-for-granted realities and accoutrements of everyday life" (Berger and Luckmann, 1967: 119). The cultural recipes for correctly dying are provided by the legitimations of death which in turn sustain the social order. The medical act of healing obviates the threat that death brings to the social order. Furthermore, as a concomitant of this, most recent studies in death and dying[1] have tended to emphasise that dying as a total experience is at variance with the hospital emphasis on the disease process and the diseased organ, where the termination of life is viewed as failure on the part of the personnel involved. The Death System (the totality of words, actions and attitudes concerning death, typified in the Ars Moriendi[2] and the Egyptian Book of the Dead) is socially constructed and perpetuated, varying from age to age and generally a reflection of the current social structure and institutions.

The social history of death

The rise of widespread interest in thanatological (death and dying) concerns is a characteristic of the seventies, and not limited to those who practice cryogenics.[3] Historically, "infectious, acute diseases were an important source of death in the premodern world" (Lofland, 1978: 21), the key factor being the absence of "early diagnosis" and the "liklihood that diseases or potentially fatal conditions would be 'identified' rather late in the dying-to-death trajectory". The modern shift in the social pattern of dying meant that:

> not only was the family as a socially-supporting institution altered by the impact of industrialisation and urbanisation, mortality and morbidity rates shifted such that death in childhood sharply declined whereas dying by lingering, chronic illness showed a remarkable rise.
> (Benoliel, 1978: 6)

Illich (1977: 179-214), following Aries, distinguished six stages of "natural death": a) the fifteenth-century "dance of the dead"; b) the "Dance of Death"; c) "Bourgois Death"; d) clinical death; e) medical intervention; f) intensive care death. Illich utilises the concept of industrialisation to explain the perceived evils of modern society, particularly medical practice and the depersonalisation of dying. But whereas Illich is essentially an advocate of changed lifestyles others, such as Navarro, would argue that this is basically a conservative approach and that the real solution lies in political change. Hence:

> the greatest potential for improving the health of our citizens is not primarily through changes in the behaviour of individuals, but primarily through changes in the patterns of control, structures and behaviour of our economic and political system.
> (Navarro, 1975: 361)

Society's attitudes to death consist of normative responses to death and the dying, and are calculated to function as constraining mechanisms. Historically, Aries (1974) has typified the attitude of the twelfth century as "tamed death" when the dying embraced their fate as an inevitable aspect of life. The Middle Ages regarded death as familiar and public and its catechism and ritual was firmly rooted or located in the family setting. In the twentieth century's attitude death is hygenic and unobtrusively private. On average, for example, an American experiences a family bereavement once every twenty years (Dumont and Foss, 1972). The avoidance of death, where death is forbidden, manifested itself in "the displacement of the site of death. One no longer died at home in the bosom of one's family, but in hospital, alone" (Aries, 1974: 87). Thus "Death is no longer an occasion of a ritual ceremony...Death is a technical phenomenon obtained by a cessation of care...determined in a more or less avowed way by a decision of the doctor and the hospital team" (Aries, 1974: 88).

That a bureaucratic method of dealing with death is currently practised should not, as Blauner (1966: 384) indicates, really surprise anyone. It is, after all, the western world's "characteristic form of social structure" rather than a manacious plot, as Illich might express it, on behalf of the Medical Establishment. Nevertheless, depersonalised, institutionalised death is concomitantly a shift in focus, away from individual decision-making in life and death:

> Dying in the nursing home meant removal from a life-stimulating environment and loss of personal control over intimate decisions to a company of strangers. The technologising of death increased the dehumanising outcomes for dying patients ...
> (Benoliel, 1978: 11)

Lofland quite rightly points out that a combination of factors such as a) a reaction against impersonalism, b) the rise in chronic illnesses, and c) an early identification of the soon-to-be-dead, manifests itself among a number of individuals and groups in an attempt to exert an existential <u>choice</u> as to the form and meaning that death will take:

> It is only very recently that any significant portion of human beings find themselves "dying" for a long enough period of time for the issue of how best to go about it meaningfully to be raised. That a new role possibility - that a new "core category" around which identity may be formed - has emerged and that this new role possibility is unencumbered by much historic structuring, engenders action problems for those who take it on.
> (Lofland, 1978: 35)

The process of becoming dying

Berger and Luckmann stated that:

> Life expectency varies with social location... there is considerable discrepancy between the life expectencies of lower-class and upper-class individuals. Furthermore, both the incidence

> and the character of pathology vary with social location. Lower-class individuals are ill more frequently than upper-class individuals, in addition they have different illnesses. In other words, society determines how long and in what manner the individual organism shall live.
> (Berger and Luckmann, 1967: 181)

Death is socially constructed and appears at least to be constrained by a Durkheimian cohesiveness or social integration (Seligman, 1974). Thus Phillips and Feldman (1973) found that deaths were less frequent before significant ceremonial occasions. And Alderson (1975: 151), utilising mortality data for England and Wales for 1972, found that persons over seventy-five displayed a consistent trend in deaths "with an excess in birth month and the following three months". This suggested the hypothesis that in the elderly "a birthday may in some subtle way influence the general morale of an individual". Finally, Engel (1971: 771) mentions that sudden death follows a precipitating life situation.

Studies of healthy populations have suggested 91%-92% have bodily disorders requiring treatment (Pearse and Crocker: 1944; Hinkle: 1960; Schenthel: 1960). The presentation of perceived symptoms to a doctor takes place in a complex system of social life which gives "meaning" to the potentially sickly facets of existence. Thus Zola states that "Instead of it being a relatively infrequent or abnormal phenomenon, the empirical reality may be that illness, defined as the presence of clinically serious symptoms, is the statistical <u>norm</u>"[4] (Freidson, 1972: 392), which is also suggested by Mechanic's (1968: 149) view that "being ill is not an unusual state of being, and much illness is ignored and tolerated". Twentieth century man has at his disposal a highly technical medical machine able to diagnose his symptoms which, combined with a cultural acceptance of entry to the stage of the "chronic sick" and treatment capable of prolonging life, produces a "monstrous regiment" of the soon-to-be-dead.

Validation of the status dying

While Scheff (1963) maintains that there is considerable evidence that physicians facilitate illness after ostensibly minor symptoms are diagnosed, it is probably not the case that this is so of dying itself. Apple (1960), Mechanic (1962), Gordon (1966), Kosa and Robertson (1969), Zborowski (1969), Robinson (1971) and Twaddle (1977) have written extensively on the sick role, some of them suggesting that it be abandoned entirely as a valid concept. The grounds for rejection of Parson's analysis are generally that: (1) the sick role lacks uniformity when applied to a variety of social groups and individuals; (2) the sick role is more appropriate to acute illness; (3) Parson's view of patient-physician interaction is traditional and one-to-one; (4) Parson's model has a middle-class orientation; (5) it does not take into account non-western societies. A radical extension of the sick role views it as providing:

> a convenient mechanism of social control by which institutions can allow deviant behaviour within carefully controlled limits. Because the sick role reduces potential opposition directed against the institutional structure itself, it fosters institutional stability.
> (Waitzkin and Waterman, 1974: 65)

Nevertheless, Parson's sick role formulation offers useful insights into the sickness career. If we start at the "being well" state there are points which are culturally defined as "not being well". Generally these points consist of (1) not feeling well; (2) inability to perform some task; (3) other more important symptoms (Baumann, 1961). Thus it appears that individuals first realise that they are sick when their symptoms interfere with their daily lives, in particular their day-to-day activities, when their tolerance threshold breaks down, and when their symptoms cannot be located in the realm of the familiar. Friedson's (1961) "lay referral system" is but the first step before the symptoms are professionally legitimated. One of the alternative routes to "getting well" is dying, a status validated by the following conditions:

1. recognition and validation by physicians;
2. impairment of normal activities;
3. chronic rather than acute symptoms;
4. uncertain or poor prognosis.

The contemporary pariah

Aries's analysis of the changing attitudes towards death begins with the sparing of the sick person the truth of his own death and ends with the initiative of death being placed in the hands of the doctor and the hospital team, away from the family. Consequently, those who control death attempt to elicit "an acceptable style of living while dying" (Aries, 1974: Gorer, 1965).[5] Blauner (1966) mentions that "many writers have commented on the tendency of relatives to avoid the subject of death with the terminally ill" and that one way of combating this meaninglessness is by adopting Glaser and Strauss's (1965) "open awareness context" where both physicians and patient express an acknowledged awareness of the terminal situation. Contemporary man "hides his death as he hides his sex, as he hides his excrements" (Morin, 1951: 331). Even in terms of physical plant "the hospital morgue is best located on the ground floor and placed in an area inaccessible to the general public. It is important that the unit have a suitable exit leading onto a private loading platform which is concealed from hospital patients and the public" (Owen, 1962: 304).

The negotiation of the terminal situation

Blauner (1966) writes about the dying, especially those from a higher social status, as being "in a better position to negotiate certain aspects of the terminal situation". Again, he mentions that:

> The dying patient in the hospital is subject to the kinds of alienation experienced by persons in other situations in bureaucratic organistions. Because doctors avoid the terminally ill, and nurses and relatives are rarely able to talk about death, he suffers psychic isolation. He experiences a sense of meaninglessness because he is typically kept unaware of the course of his disease and his impending fate...He is powerless in that the medical staff and the hospital organisation tend to program his death in keeping with their organisational and professional needs; control over one's death seems to be even more difficult to achieve than control over one's life in our society.
> (Blauner, 1966: 89)

Pacing and styling the dying role

Three characteristics of the dying role mark it out as singular. Firstly, it is a transitional status passage; secondly, it is generally irreversible; and lastly, it is a non-graduating condition in the sense that no one returns to bear tidings as to the quality of the experience. There are also clearly defined rules of entry such that one has to be admitted by the gatekeepers of death and the "actors have to accept the admission as accurate" (Lofland, 1978: 45).[7] Why the label "dying" should be ascribed to some and not others in our society is socially determined and prescribed, and the label "dying" is imputed or applied depending on the dying person's location in a variety of social structures and the labeller's involvement in a mesh of organisational, interactional and professional situations (Sudnow, 1967: 68-72). Some, such as Kühbler-Ross, have postulated phases of the dying role or "stages of death": denial, anger and rage, bargaining, depression and acceptance, although these only describe the possible course of reaction of the soon-to-be-dead to their impending fate.[8] Noyes's (1974) model of the death experience postulated three stages, namely a) resistance; b) review; and c) transcendence.

Lofland, in her study, further elaborates on the process of styling to include "cultural scripts" the function of "which provide some specification for how one is to act when one is in the category" (Lofland, 1978: 48). Other aspects include space (what proportion of the individual's identity is to be labelled "dying"), population (whether the dying role is to be played in a privatised or communal setting), knowledge (who shares the knowledge of the individual's admission to the dying role) and stance (the verve with which one enacts the role). Further, the degree of panache to which we can control or manage the dying role is determined by four "external" conditions, namely, "(1) the disease process, (2) the social organisation and culture of medical practice, (3) available resources, and (4) surrounding others" (Lofland, 1978: 58).

The biomedical model

The growth of the biomedical model manifests itself in a number of ways:

> objectification of the doctor-patient relationship through formalization of the patient history and physical examination; use of technical procedures to assist in diagnosis; utilization of machines in treatment; clinical experimentation; and the appearance of multiple fields of specialized medical practice...According to (Engel's) argument, the biomedical model of disease has acquired the status of dogma such that all members of society are influenced to a greater or lesser extent by a belief system that defines disease in terms of biomedical and physical deviations from established norms and a reductionist approach to the problem of defining illness.
> (Benoliel, 1978: 5; Engel, 1960, 1971)

Perhaps Illich puts it more graphically:

> ...the man best protected against setting the stage for his own dying is the sick person in the critical condition...The medicalization of society has brought the epoch of natural death to an end. Western man has lost the right to preside at his act of dying...Mechanical death has conquered and destroyed all other deaths.
> (Illich, 1977: 210)

Arluke (1978: 109) states that "death can challenge the efficacy of the individual practitioner as well as the profession as a whole". The medical personnel buttress themselves against the instrusion of a death event by "concentrating the work of review on the post-mortem course of patients" and its significance is "diminished by directing instructional interest on the events that lead up to death" (Arluke, 1978: 114). Physicians not only determine the circumstances of death but the exact moment at which it occurs. The routinisation of death in the hospital setting is a process of disengagement from a perspective of the dying as a social object. Staff enter into a series of avoidance procedures which have been documented by Sudnow (1967). There is some evidence that the medical model enables the student physician to become socialised to death:

> ...the clinician learns to view dying patients not as people with feelings, but as medical entities, specimens, or objects of scientific interest. By adopting a scientific frame of mind, utilized so effectively in their previous work with dead bodies, clinicians can effectively avoid the uncomfortable inner feelings which occur when they are exposed to dying patients. (Coombs and Powers, 1975: 258)

The Mask of Professionalism, with its special language, attire, cynicism, material objects (e.g., stethoscopes), ritualised and impersonal action, and hospital routine, are addenda to the socialised perspectives on death (Bowers et al, 1976: 52-73). Death has become shrouded, softened, in a bureaucratic system because the "death specialists" have taken over various handling procedures (Blauner, 1966). However, an inevitable dichotomy exists between the rational-bureaucratic model and the affective implications of grief and mourning. Furthermore, hospital personnel can become "acclimatized" to death and dying, usually when "Death Saturation" occurs and the staff become overloaded with too many deaths (Pattison, 1977: 245-267).

Thus Benoliel speaks of a "counterculture against impersonal death", probably beginning with Feifel's (1959) work, culminating with Kühbler-Ross (1969). Death and Dying courses not only appear in the high school curriculum but also in newspapers, notably in California.[9] Lofland speaks of the same counterculture as the "Happy Death Movement":

> ...the individuals, organisations, and activities which are the movement are concerned with promoting change in American society with regard to its beliefs about death and dying, its emotional responses to death and dying, and its legal and normative practices relative to death and dying. (1978: 77)

Listed among the agencies pledged to combating the shift in control, from the individual to a medicalised, technologised, impersonal mode of medical engineering, are such groups and organisations as the "Euthanasia Education Council, the Foundation of Thanatology, Ars Moriendi, the Forum for Death Education and Counselling, Shanti, Threshold, Inc.[10] Bereavement Outreach, the Widow-to-Widow Program, Hospice of Marin, California,[11] Hospice, Inc., of New Haven, Connecticut, and Compassionate Friends..."[12] The informal death talk therapy in such movements follows the well-heeled pattern established by Alcoholics Anonymous and Recovery, Inc (Jones, 1970: 1975 See Chapter Eight): "My name is Kevin and I am an alcoholic. I have been drinking for as long as I can remember", or "My name is Pat and I am an ex-mental patient. Last week I couldn't leave the house without palpitations". One description is as follows:

> Each meeting opens with an affirmation of personal tragedies in the style of Alcoholics Anonymous: "My name is Ernie Freireich and I am a bereaved parent. I lost my son Mark in an automobile accident." Each lasts as long as members want to air feelings and think they can provide mutual support.
>
> "Only in such settings," explained Freireich, whose 16-year-old son was killed two years ago, "do you feel able to talk freely with another person who's been down the same street you've been down, without having judgement passed on you." Knowing that their listeners understand the participants feel free to express the depths of their pain and bitterness.
> (Seligman and Agrest, 1977: 89)[13]

Self-help groups

Groups concerned with self-help have mushroomed dramatically in the last few years, particularly in North America, developing, generally, from "grass-roots origins, their faith in group dynamics as a source of individual change and their reliance on the sharing of common experiences rather than professional expertise..." (Lusky and Ingman, 1979).[14] Some have classified the self-help group occurrence as a "collective response to an irrational and fragmented society" (Katz and Bender, 1976: 37-38). Such groups possess the following attributes: (1) they involve other persons on a "small-group", "face-to-face" basis; (2) they arise usually spontaneously; (3) they are sometimes <u>ad hoc</u> or short-lived; (4) they vary in functions and characteristics; (5) they involve personal participation; (6) they emphasise the <u>meaning</u> of participation; (7) they supply a reference group; (8) they require some form of action; (9) they usually begin from a condition of powerlessness.[15]

Developmentally self-help groups emerge from social movements, emphasising <u>change</u> as their main orientation. The characteristics of self-help groups appear to approximate to two of Blumer's (1951: 199-220) categories of social movements, namely that they are <u>specific</u> and <u>expressive</u> rather than general. Katz and Bender (1976), using a primary focus framework, construct a typology of self-help groups, namely: (1) groups whose primary focus is on self-fulfilment or personal growth; (2) groups whose focus is on social advocacy; (3) groups focusing primarily on the creation of alternative patterns for living; (4) "outcast haven" or "rock bottom" groups (1976: 37-38). Cressey (1955) adds a further dimension, namely rehabilitation, while Sagarin (1967) supplies "conforming" and "reforming", "inner-focused" and "outer-focused".

Happy Death Movements, as self-help groups, arose principally as a reaction against the bureaucratisation of death:

> In the main, movement proponents and sympathisers have not found these arrangements satisfactory. They have argued...that hospitals' emphasis on 'cure' makes them poor settings for prolonged dying: that the needs of the dying are submerged to the routine of instrumental care...
> (Lofland, 1978: 84)[16]

The classification of self-help groups

Primary Focus	Inner Directed	Outer Directed	Conforming	Reforming
Self-orientated	Recovery, Inc.	-	Yes	Yes
Socially-orientated	-	Welfare Rights Organisations	-	Yes
Alternative-orientated	Gay Liberation	Women's Liberation	-	Yes
Omega or "Rock Bottom-orientated	X-Kaley		Yes	Yes

There seems little doubt that the reaction against the routinisation and bureaucratisation of death has manifested itself in a quest for autonomy, in a desire for the power of negotiation to determine one's death. Crane (1975) sees such a reaction as forming a loose social movement which has two primary goals, a) improvement of the quality of interaction between, and with, the dying; b) seeking a legislative right to refuse treatment.

Rational - buraucratic aspects of terminal care

Kühbler-Ross's emphasis has been on death as the final stage of growth, a legacy of an essentially developmental model (Charmaz, 1976),[17] in the belief that "When you are dying, if you are fortunate enough to have some prior warning...you get your final chance to grow, to become more truly who you really are, to become more fully human". The inexorable drift away from the humanisation of death towards a rational-bureaucratic management where the dying are technologically processed is noticed by a large number of writers. Thus Reiser (1977: 229) states:

> Machines...direct the attention of both doctor and patient to the measurable aspects of illness but away from the "human" factors that are at least equally important...the problems of illness

> hatched from beliefs, illusions, values and other facets of cultural and mental life...are as forceful and significant as the biological problem of illness approached through technology and scientific learning.

Technology's "prolonged assault on death" can be evidenced from four examples: a) we will compare the numbers of those dying in hospitals over the last ten years and if it is the case that a greater percentage are now dying in a hospital situation, and hospitals are highly technologised institutions, then the dying act as itself becoming more technologised; b) we will cite evidence from a contemporary sociological account of hospital dying; c) we will cite evidence from a dying "timetable" and a dying "event"; and d) we will record a personal description of one woman's death by her son.

A dying "timetable" could be described as follows:

8,20 p.m.	Admitted to the hospital.
11,15 p.m.	Emergency operation for cerebral damage following a road accident which caused a posterior fossa fracture resulting in swelling in the occiput and compression.
2,10 a.m.	Operation complete; intravenous feeding.
6,30 a.m.	Machine-controlled life by being placed on respirator.
7,15 a.m.	Prognosis for recovery nil.
6,00 p.m.	Neurological administration of electroencephalogram reveals no cortical activity or evidence of viability.
10,00 p.m.	Operating room. Normal body temperature, pulse, blood pressure and respiration.
11,00 p.m.	Removal of respirator.
11,15 p.m.	Patient pronounced dead. Removal of heart and kidneys for recipients.[18]

This is clearly a case of dying without knowing that one is dying as the patient, after the initial operation, has no perceived awareness. Nevertheless, the prominence of technology is obvious, and the dying process can be compared with the account below:

> The woman lay half-naked on the bed, hooked up on tubes, a tracheostomy, and a respirator, staring desperately around the room. The psychiatrist's first impulse was to cover her with a bed-sheet, but a nurse appeared and said, "Don't bother - she will push it off again in a minute." The psychiatrist approached the patient, who

> took her hand and pointed to the ceiling. The psychiatrist looked up and asked if the light was bothering her...when the psychiatrist asked for the light to be switched off, however, the nurse reminded her of the rules and regulations of the intensive treatment unit, which required that the light stay on. Then the psychiatrist asked for a chair for the mother to sit with her daughter. She was told that they could not give her a chair anymore because during the previous visit the mother had stayed more than five minutes. The woman died eight hours after the physicians had informed the parents of her imminent death - she died with the room light in her eyes, the tubes in her mouth and veins, and the parents sitting outside in the waiting room. (Vaatch, 1977: 319-320)

Sudnow's distinction between "clinical death" (the appearance of "death signs"), "biological death" (the cessation of cellular activity) and "social death" (the patient is treated as a corpse) is of interest here (1967: 74):

> A nurse on duty with a woman she explained was "dying" was observed to spend some two or three minutes trying to close the woman's eyelids...When questioned about what she had been doing, she reported that a patient's eyelids are always closed after death, so that the body will resemble a sleeping person. After death, however, she reported, it was more difficult to accomplish a complete lid closure...she always tried...to close them before death; while the eyes are still elastic they are more easily manipulated. This allowed ward personnel to more quickly wrap the body...

Most of Sudnow's observations were of sedated, comatose, deaths. Indeed, of 200 deaths observed, no "deaths were of the Hollywood version, wherein the person's last sentence is interrupted by his final breath" (89). He cites a passage in support of his observations, to the effect that "the classical deathbed scene, with its loving partings and solemn last words, is practically a thing of the past; in its stead is a sedated, comatose, betubed object, manipulated and subconscious, if not subhuman" (Fletcher, 1960: 141).

While hospitalised deaths in America have been variously given as 73,1% (Lerner, 1970) and 50% (Morison, 1973), in Britain the percentage of deaths that occurred in hospitals in 1967 and 1977 were as follows:

Hospital deaths as a percentage of total deaths in Great Britain in 1967 and 1977

		Total Deaths	Hospital Deaths	%
1967	M	307,732	165,706	53,84
	F	294,307	163,014	55,38
1977	M	321,053	186,760	58,17
	F	317,169	194,250	61,24

Source: Office of Population Censuses and Surveys, 1979.

This shows an increase of 4,32% for men in ten years, and 5,85% for women. The total of deaths by all causes according to type of institution for England and Wales for 1967 and 1972 shows the same pattern of increase:

Total deaths by all causes, and sex, according to type of institution, in England and Wales, 1967 and 1972

	Total Deaths		Psychiatric Hospitals[i]		Other Hos-Hospitals[ii]etc		Home[iii]	
	M	F	M	F	M	F	M	F
1967	277,178	265,338	6,728	9,980	149,085	149,623	121,365	105,735
1972	300,389	291,500	6,150	9,294	173,016	179,974	121,223	102,232

Source: Registrar General's Statistical Review of England and Wales, 1967 and 1972, Part 1, Appendix H.4.

Notes: [i] including N.H.S. and non-N.H.S.
[ii] including other institutions, both N.H.S. and non-N.H.S.
[iii] at the deceased person's own home, other private houses and other places.

In terms of percentages relating to two categories, viz. institutional and non-institutional deaths, we have:

Institutional and non-institutional deaths in England and Wales expressed in percentages, 1967 and 1972

	Institutional Death		Non-Institutional Death	
	M	F	M	F
1967	56,21	60,15	43,78	39,84
1972	59,60	64,92	40,35	35,07

Thus in 1972 the percentage of deaths in England and Wales that occurred in institutions was 59,60% for men, an increase of 3,39% from 1967, and 64,92% for women, an increase of 4,77%. Undoubtedly, therefore, the increasing tendency is towards a highly technologised death in an institutional, rather than a domestic, setting.

Finally, a personal account of one woman's death by her son further emphasises the impersonal aspect:

> 7 p.m. Mother took ill soon after a light tea. The doctor visited her at 11 a.m. and, after examining her, suggested that she be kept on the oxygen. As soon as he had gone she lost her speech. I presumed that she had had a stroke. I telephoned the ambulance which arrived in a short time. The men said that my mother would be admitted to the new Royal County Hospital but, half way there, they were informed by radio that there were no beds available and that they were to proceed to Shire Hospital. The doctor at Shire said, "Your mother has had a stoke but isn't ill enough to be detained in hospital. What's more we have no spare beds and so we'll send her home as soon as an ambulance is available." I waited until 9 p.m. She was kept in a cubicle for six hours without my seeing her again that night. I went home to await her return and shortly afterwards the hospital rang to say that my mother had had another slight stroke and that they were admitting her to a ward. It later transpired that she had pneumonia. She died two days later...I must add that the doctor was frightfully arrogant. I had been nursing my mother for the last two years, day and night. I slept on the couch so that I could minister to her bodily functions and clean her, and to give her oxygen frequently throughout the night. I know my mother had to die sometime, but not to be treated in such a despicable manner before doing so, after all my devoted attention.[20]

Making today count

The increased rational-bureaucratic management of death invariably results in episodes of dehumanisation such as those described above. The counterculture against such technological intrusion is well represented by a group called Making Today Count (MTC), Inc., which has 136 Chapters across the United States. The founder, Orville Eugene Kelly, gives an account of his own cancer and the subsequent beginning of the movement in Make Today Count,[21] published in 1975. Kelly's own reaction to the diagnosis of malignant lymphoma closely parallel Kübler-Ross's "stages of death". He writes: "I was terrified and I was filled with fury" and "I wanted to know why I was being shortchanged".[22] Following his newspaper account of his illness the initial MTC meeting of eighteen cancer patients and their families was held at the Elks Club in Burlington, Ohio. Kelly's primary aim was that:

> Cancer has to be faced. Death has to be talked about openly...Only that way can we give our-

selves a chance to take pleasure in our daily lives, and give our lives some sense of meaning and accomplishment.

Organisation

Membership of MTC increased rapidly.[24] The format of the meetings is similar to that of A.A. and other self-help groups - namely a sharing of experiences, the appearance of guest speakers, and the ability of the member to make a decision that they want to be helped. Information is shared by the production of a monthly newsletter.[25,26]

Beliefs and practices

Kelly (1975: 127) claims that:

> The entire thrust of MTC, both as a philosophy and as a movement, is to improve the quality of life for people whose terminal illness is a real and personal matter - and those are, of course, the terminally ill and their families.

He is convinced that most people fear death because they don't talk about it freely:

> Probably because I live so close to the knowledge of my own death, I have become particularly sensitive to the ways people try to deny death, and... that it leads ultimately to a denial of life itself (129).

The emergence of this new "core category", the chronically dying, is a relatively new phenomenon, as Lofland points out above, and, as a reaction against the technologising of death and its concomitant dehumanisation, the soon-to-be-dead are encouraged, in groups such as these, to make a conscious choice as to the form that their remaining days will take. By such conscious measures the dying obviate what Blauner describes as the psychic alienation so commonly experienced: in sum they establish control over their dying.

The major problems accompanying cancer are largely emotional. However, the success of MTC, as reflected in its phenomenol growth, is indicative of many things:

> the need of the terminally ill to share their problems and to learn from one another; the growing willingness of our society to deal with the problems of illness and death; our growing unwillingness to tolerate cover-ups, whether political or medical. It also reflects, I think, another development: a growing independence among patients and their families...We are no longer content to be treated like children...the

best treatment for us as human beings...is compassionate honesty (146).

Death can be more tolerable if it can be publicly discussed and society can be made to change its attitudes. MTC thus is comprised of individuals the major part of whose identity is labelled "dying" who are playing their roles in a communal setting in a flamboyant announcement of their dying status.

The basic tenets of the group are crystallised in its goals and philosophy.[27] MTC is not a religious organisation nor is it involved with any religious sect or denomination. This does not mean that the organisation prohibits the sharing of a variety of spiritual or religious experiences. Kelly himself is a religious person but MTC does not have a mandatory belief to which members have to subscribe. The organisation is emphatic that it does not become involved in any recommendations regarding certain kinds of treatment, and neither does it provide ancillary services such as nursing and homemaking.

The dynamics of anticipatory death

MTC is not the only self-help group for the terminally ill. For example, Reach for Recovery is composed of women who have had mastectomies and who are committed to helping others.[28] Parents' Sharing Group, while not composed of the soon-to-be-dead, "is a group process designed to help parents of children suffering from leukemia cope with some of the human threats that accompany diagnosis, treatment and hospitalisation of the disease" (Kerney, 1974: 316-319). The usual type of group is composed "of three to eight mothers, and is fluid in nature, with new persons coming in and others dropping out".

The dynamics of all such groups, however, are concerned not only with anticipated death but also with anticipatory grief. Such grief, as documented by Aldrich (1974), has something to teach us concerning the quality of experience felt by the soon-to-be-dead. For one thing, in comparison with conventional grief anticipatory grief accelerates rather than decelerates. There is a certain amount of unresolved <u>ambivalence</u> attached to it, also, which serves merely to increase the prevalence of <u>denial</u>.

Certain characteristics are common to all aspects of the dying role:

> Over an extended time, therefore, the family members may (1) experience depression, (2) feel a heightened concern for the ill member, (3) rehearse his death, and (4) attempt to adjust to the various consequences of it. By the time the death occurs the family will, to the extent that they have anticipated the death or dissi-

pated their grief, display little or no emotion. (Fulton and Fulton, 197)

The formulation of the personal and social meaning of an anticipated death involves not only questions of the type asked by Kelly but also coming to terms with the significance of certain values in the life that one has hitherto led. It includes, also, a certain adjustment to the personal time-space dimension because "anticipatory grief involves a profound sense of loss of the future" (Ramshorn, 1974: 248). Neither does anticipatory grief, in the usually short period it is with us, allow the soon-to-be-dead and his family to revamp their emotional orientations. "Anticipatory grief is a period of intense involvement, which requires suspension of normal roles within the family, work and community systems." (Ibid, 249)

Finally, professional treatment of the sick relates acutely not only to the patient's physiological status but also to his social situation (Crane, 1975: 68-84).[29] The terminally ill fluctuate between seeking treatment actively or not, and the attitude of the patient is an important factor in the physician's decision making process, apart from the evidence that a number of patients have difficulty in conveying their attitudes to the medical staff. The family's attitude is regarded with some ambivalence by the physician, with the exception perhaps of the treatment of an unsalvageable patient. A number of variables influence the decision process. High status patients are differentially treated as against low status patients - but not if they are unsalvageable. Deviants such as prostitutes or addicts, contrary to Sudnow's experience, did not appear in the Crane study to be discriminated against. Finally, age seems to influence the degree of active treatment so that those showing serious illness at the middle of the life-cycle were more likely to be accorded suitable treatment than those at the end of the life-cycle.

We have argued that an increasing depersonalisation of death, arising from a highly technological approach, has resulted in a quest for autonomy manifested in some form of personal control of the dying event. This reaction, displayed in concrete terms in the growing number of dying groups such as MTC, arises out of an inevitable dichotomy between the rational-bureaucratic model and the affective implications of the dying role. At present self-help groups for the terminally ill coexist alongside the medical profession in an uneasy truce. Their very existence probably implies the latter's laxity in supplying the affective dimension to merely physical treatment. The work of Cicely Saunders in London, Brown in Minnesota, Twycross in Oxford, Gusterson in Sussex, Hadlock in Orlando, Florida, Lamers at Marin in California, Mount in Montreal, are just some of those active in the field of terminal care within the medical profession. Currently there are about forty hospices in England with others emerging in North America and Europe. Movements such as MTC have an important lay contribution to make because they see their programmes from the perspective of the dying patient. They are singular movements emphasising a positive approach for the soon-to-be-dead, creating a situation where the latter can negotiate their own identity.

NOTES TO APPENDIX "B"

[1] These include work by Sudnow (1967), Kübler-Ross (1969), Vernon (1974), Fulton (1965), Glaser and Strauss (1965) and Mauksch (1974). Gustafson (1972) for example, applied the concept of career to the study of a nursing home for the terminally ill where "For the aged patient, the passage of days itself brings him noticeably closer to the end of the road". Sudnow analyses the categories of hospital life as they are seen to be "constituted by the practices of the hospital personnel as they negage in their daily routinised interactions within an organisational milieu". Glaser and Strauss discuss the dying <u>trajectory</u> during which the person dying exhibits a career or status passage which has <u>duration</u> and <u>shape</u>: "(1) The patient is defined as dying. (2) Staff and family then make preparations for his death, as he may do himself if he knows he is dying. (3) At some point, there seems to be 'nothing more to do' to prevent death. (4) The final descent may take weeks, or days, or merely hours, ending in (5) the 'last hours', (6) the death watch, and (7) the death itself." As a trajectory death is "almost completely unscheduled; second, the sequence of steps is <u>not institutionally prescribed</u>; and third, the actions of the various participants are only <u>partly regulated</u>".

[2] The <u>Ars Moriendi</u> was a do-it-yourself manual on the etiquette of dying. See Illich (1977: 188-190). Also Mary Catherine O'Connor, <u>The Art of Dying Well: The Development of the Ars Moriendi</u>, New York: AMS Press, 1966.

[3] See Sheskin's forthcoming book on Cryonics. The rate of morbidity and a society's interest in death is not clearly correlated. Death and dying in Puritan New England (Stannard, 1973) was as much an all-pervading concern as it had been in the Middle Ages (Cohn, 1961). The contemporary concern with death is essentially a <u>middle-class</u> involvement for the simple reason that it is this sector of society which is most aware of the differential mortality rates despite the fact that mortality data based on occupation and social class actually favours classes 1 and 11, i.e., death rates in the western world vary inversely with socio-economic status. See also Zelizer (1978), which explores the emergence of life insurance as a new form of ritual with which to face death whereby human life is converted into commodities.

[4] Zola precis Hinkle as follows: the average lower middle-class male between the ages of 20 and 45

experiences over a 20-year period approximately one life-endangering illness, 20 disabling illnesses, 200 nondisabling illnesses and a 1000 symptomatic episodes. These total 1,221 episodes over 7,305 days or a new episode every six days...In short even among a supposedly "healthy" population scarcely a day goes by wherein they would not be able to report a symptomatic experience (Freidson, 1972: 392).

[5] Aries gives Gorer credit for pioneering modern death sociology.

[6] Quoted in Sudnow (1967: 51).

[7] It is not true, however, that none purport to describe death experiences (Lofland, op.cit, p.94). One such study, outlined in Science Digest, Vol.81, No.2, February 1977, pp.71-72, by John White, describes a forthcoming study by Karlis Osis and Erlendur Harolsson, At the Hour of Death (New York: Avon). 50,000 dying in the USA and India were studied and among the conclusions were the following: (1) elevation of mood; (2) death-bed visions either of a person such as a relative or a religious apparition, or a vision of another reality.

[8] For criticism of Kühbler-Ross's "stages of dying" see Gullo et al in Schoenberg (1974), and Lofland (1978: 48). For criticism of Kühbler-Ross's mysticism see Lofland pp.103-104, notes 18 and 19.

[19] Kelly (1975) mentions 1,500 courses on death and dying in the USA. The Observer featured an article by John Barnes entitled "Lessons in Death":

> Death is coming into its own in the elementary schools of America...teenagers meet with terminally ill patients at a local hospice, where lives are measured in weeks and months...students are learning the mechanics of the American free enterprise system by making and selling a product that's always in demand - coffins.

[10] Lofland mentions that this particular group rents out dying companions.

[11] For a description of the work at Marin see Stoddard (1978: 49-69).

[12] See also Renee Fox and Diane Crane (forthcoming) The Death and Dying Movement. According to Lofland the social composition of the groups is "white, relatively affluent, both straight and counterculture in style, and, unsurprisingly heavily representative of such occupations as physician, nurse,

clergyman, social workers, psychiatrist, psychologist and counsellor"(77).

[13] See Jones (1970; 1975); Sagarin (1969; 211-232); Katz and Bender (1976), and especially Lee (1971: 42-48).

[14] This article contains a fairly comprehensive account of The Fellowship, an alternative to Alcoholics Anonymous, which appeals mainly to those equipped with less verbal skills.

[15] Katz and Bender (1976: 3-12). The historical development of self-help groups can be traced from the Freemasons of the Middle Ages through the English Friendly Societies to producer-consumber cooperatives, the trade unions and mutual aid.

[16] Concomitant alternative models, according to Lofland, are: (1) the creation of a "dying" place, and (2) the creation of a "home death". Both emerged from the work of Cicely Saunders, founder of Lodnon's St. Christopher's Hospice. See Kron (1976) and Saunders (1977a; 1977b).

[17] Quoted in Lofland, op.cit., p.48.

[18] Adapted from Veatch, 1977: 317-347.

[19] After Kühbler-Ross.

[20] Extracts from a diary, dated 11.1.79, kindly provided by Mr. A.A. Phillips.

[21] Written in conjunction with Randall Becker, New York: Delacorte Press. My thanks are due to officials of MTC who provided me with access to much of the following information.

[22] <u>Make Today Count</u>, p.11.

[23] Ibid. p.21.

[24] Organisation and structure of MTC groups approximated well to Lofland's observations quotes in note [12] above:

> ...we needed a central clearing house...we applied for a federal charter as a non-profit, tax-exempt corporation in the State of Iowa...According to its charter, Making Today Count is made up of two groups of people. Patients and their families or survivors - and by patients we mean not only cancer victims, whether their prognosis is favourable or unfavourable, but also victims of other serious incurable

diseases - are the central core, the members proper. In addition we have associate members - physicians, nurses, nursing students, ministers, social workers, college students.
(Kelly, 1975: 96-97)

[25] Containing articles such as the National Cancer Institute's report on Lactrile which appeared in full in the <u>New England Journal of Medicine</u>, September 7, 1978.

[26] <u>Making Today Count Newsletter</u>, No.18, p.1. Only 10,6% of expenses went on fundraising. Twenty-three and two tenths per cent were paid for management expenses and "the balance went to provide information services, to start new chapters, promotion of better management services, publications, annual workshops, and co-operation with other health agencies".

[27] The philosophy and goals of the movement are stated as follows: (1) to help the patient and his or her family cope with life-threatening illnesses; (2) to improve the quality of life for all persons with serious illness; (3) to identify emotional problems of life-threatening illnesses and to teach people to cope with them; (4) to promote openness and honesty in discussing and dealing with a serious illness; (5) to assist the professional in communicating and meeting the needs of the patient, family and friends, who are faced with a life-threatening problem.

[28] Death from cancer is now the second leading cause of death. The mortality incidence following a removal of a breast is still extremely high. Thus Illich (1977: 33) states: "there is little effective of most other cancers. The five year survival rate in breast cancer cases is 50%, regardless of the frequency of medical check-ups and regardless of the treatment used. Nor is there evidence that the rate differs from that among untreated women".

[29] Sudnow (1967: 103) notes that "Currently, at County, there seems to be a rather strong relationship between the age, social backgrounds, and the perceived moral character of patients and the amount of effort that is made to attempt revival when 'clinical death signs' are detected (and, for that matter, the amount of effort given to forstalling their appearance in the first place)." The "moral character" of an alcoholic, for example, brings into force a set of reservations concerning further treatment (104). There is considerable evidence in Britain that both medical and para-medical staff

have reservations concerning the care and attention "wasted" on "overdosers" and attempted suicides.

****

Alderson, M., "Relationship Between Month of Birth and Month of Death in the Elderly", British Journal of Preventative and Social Medicine, 29, 1975, 151-156.

Aldrich, C.K., "Some Dynamics of Anticipatory Grief", 1974, in Schoenberg, et al.

Apple, D., "How Laymen Define Illness", Journal of Health and Social Behaviour, Fall, 1960, 219-225.

Aries, P., "The Reversal of Death: Changes in Attitudes towards Death in Western Society", in Stannard, D., ed., Death in America, 1975, Pennsylvania: University of Pennsylvania Press, 134-158.

Aries, P., Western Attitudes towards Death, 1974, Baltimore: John Hopkins University Press.

Arluke, A., "Social Control Rituals in Medicine", in Dingwall, R., et al. eds, Health Care and Health Knowledge, 1978, London: Croom Helm, 107-127.

Baumann, B., "Diversities in Conceptions of Health and Physical Fitness", Journal of Health and Social Behaviour, 2, Spring, 1961, 39-46.

Benoliel, J.Q., "The Changing Social Context for Life and Death Decisions", Essence, 1978, 5-14.

Berger, P., and Luckman, T., The Social Construction of Reality, 1967, New York: Anchor.

Blauner, R., "Death and Social Structure", Psychiatry, 24, 1966, 378-394.

Blumer, H., New Outline of the Principles of Sociology, ed., Lee, A.M., 1951, New York: Barnes and Noble.

Bowers, M.K., Jackson, E.N., Knight, J.A., Le Shan, L., Counselling the Dying, 1976, New York: Jason Aronson.

Charmaz, K.C., "A Symbolic Interactionist Critique of Kühbler-Ross's Stages of Dying", Paper presented to the American Sociological Association, New York, September, 1976.

Cockerham, W.C., Medical Sociology, 1978, New Jersey: Prentice Hall.

Cohn, N., The Pursuit of the Millenium, 1961, New York: Harper and Row.

Coombs, R.H., and Powers, P.S., "Socialization for Death: the Physician's Role", Urban Life, 4, 1975, 250-271.

Crane, D., The Sanctity of Social Life: Physician's Treatment of Critically Ill Patients, 1975, New York: Russell Sage Foundation.

Cressey, D.R., "Changing Criminals: the Application of the Theory of Differential Association", American Journal of Sociology, 61, 1955, 116-120.

Douglas, M., Natural Symbols, 1973, Harmondsworth: Penguin Books.

Dumont, R.G., and Foss, D.C., The American View of Death, 1972, Cambridge, Mass.: Schenkman.

Engel, G., "A Unified Concept of Health and Disease", Health and Disease Prevention in Biology and Medicine, 3, 1960, 450-485.

Engel, G., "Sudden and Rapid Death During Psychological Stress: Folklore or Folkwisdom?", Annals of International Medicine, 1971, 74, 771.

Engel, G., "The Need for a New Medical Model: a Challenge for Biomedicine", Science, 196, 1977, 129-136.

Feifel, H., The Meaning of Death, 1959, New York: McGraw Hill.

Fletcher, J., "The Patient's Right to Die", Harper's, 221, 141, 1960.

Fox, R., and Crane, D., The Death and Dying Movement, (forthcoming).

Freidson, E., Patients' Views of Medical Practice, 1961, New York: Russell Sage Foundation.

Freidson, E., Medical Men and their Work, 1972, Chicago: Aldine.

Fulton, R., ed. Death and Identity, 1965, New York: Wiley.

Fulton, R., and Fulton, J., "Anticipatory Grief: a Psychosocial Aspect of Terminal Care", in Schoenberg, et al. eds. 1974.

Glaser, B., and Strauss, A., Awareness of Dying, 1965, Chicago: Aldine.

Gordon, G., Role Theory and Illness, 1966, New Haven, Conn: College and University Press.

Gorer, G., Death, Grief and Mourning, 1965, New York: Doubleday.

Gullo, S.V., Cherico, D.J., Shadick, R., "Suggested Stages and Response Styles in Life-Threatening Illness" in Schoenberg, et al. eds. 1974.

Hinkle, L.E., Redmont, R., Plummer, N., and Wolff, H., "An Examination of the Relation Between Symptoms, Disability, and Serious Illness in Two Homogeneous Groups of Men and Women", American Journal of Public Health, 50, 1960, 1327-1336.

Hinton, J., Dying, 1967, London: Penguin.

Illich, I., Limits to Medicine, 1977, Harmondsworth: Pelican.

Jackson, C.O., ed., Passing: the Vision of Death in America, 1977, Westport, Connecticut: Greenwood Press.

Jones, R.K., "Sectarian Characteristics of Alcoholics Anonymous", Sociology, 4, 2, 1970, 181-195.

Jones, R.K., "Some Sectarian Characteristics of Therapeutic Groups with Special Reference to Recovery, Inc., and Neurotics Nomine", in Wallis, R., Sectarianism: Analyses of Religious and Non-Religious Sects, 1975, 190-210, London: Peter Owen.

Kastenbaum, R.J., and Aisenberg, R., The Psychology of Death, 1974, London: Duckworth.

Kastenbaum, R.J., Death, Society and Human Experience, 1977, St Louis: C.V. Mosby Company.

Katz, A.H., and Bender, E.I., The Strength in Us, 1976, New York: New Viewpoints.

Kelly, O.E., Make Today Count, 1975, New York: Delacorte Press (with Randall Becker).

Kerney, L.G., "The Ministry and a Parents' Sharing Group" in Schoenberg, B., et al, 1974.

Kleinman, A.M., "Medicine's Symbolic Reality: On a Central Problem in the Philosophy of Medicine", Inquiry, 1973, 16, 206-213.

Kosa, J., and Robertson, L., "The Social Aspects of Health and Disease", in Kosa, J., Antonovsky, A., and Zola, I., eds. Poverty and Health, 1969, Cambridge, Mass.: Harvard University Press.

Kron, J., "Designing a Better Place to Die", New York, March 1, 1976, 43-49.

Kühbler-Ross, E., On Death and Dying, 1969, New York: MacMillan.

Lee, D.T., "Therapeutic Type: Recovery, Inc." in Katz and Bender, 1976.

Lerner, M., "When, Why and Where People Die", in Brim, O., Jr. et al. eds. The Dying Patient, 1970, New York: Russell Sage Foundation.

Lofland, L.H., The Craft of Dying, 1978, Beverly Hills: Sage.

Lusky, R.A., and Ingman, S.R., "The Pros and Cons and Pitfalls of Self-Help Rehabilitation Programs", Social Science and Medicine, 13, 1979, 113-121.

Mauksch, H.O., "The Organistional Context of Dying", in Kühbler-Ross, E., ed. Death: the Final Stages of Growth, 1975, Englewood Cliffs, N.J.: Prentice-Hall, 7-24.

Mechanic, D., Medical Sociology, 1968, New York: Free Press.

Mechanic, D., "The Concept of Illness Behaviour", Journal of Chronic Diseases, 1962, 15, 189-194.

Morin, E., "L'Homme et la Mort Dans L'Histoire", 1951, Paris: Correa.

Morison, R.S., "Dying", in Life and Death and Medicine, 1973, 39-45, San Francisco: W.H. Freeman.

Navarro, V., Class Struggle, the State and Medicine, 1978, London: Martin Robertson.

Navarro, V., "The Industrialization of Fetishism or the Fetishism of Industrialization: a Critique of Ivan Illich", Social Science and Medicine, 9, 1975, 351-363.

Noyes, R., "The Experience of Dying", Psychiatry, 35, 1972, 174-184.

Osis, K., and Haroldson, E., At the Hour of Death, 1979, New York: Avon.

Owen, J.K., Modern Concepts of Hospital Administration, 1962, Philadelphia: Saunders.

Pattison, E.M., The Experience of Dying, Englewood Cliffs, N.J.: Prentice-Hall, Inc., 1977.

Pearse, I.H., and Crocker, L.H., The Peckham Experiment, 1944, London: Allen and Unwin.

Phillips, D.P., and Feldman, K.A., "A Dip in Deaths Before Ceremonial Occasions: Some New Relationships Between Sociological Integration and Mortality", American Sociological Review, 38, 1973, 678-696.

Pincus, L., Death in the Family, 1976, London: Faber and Faber.

Ramshorn, M.T., "Selected Tasks for the Dying Patient and Family Members" in Schoenberg, et al. 1972.

Reiser, S.J., Medicine and the Reign of Technology, 1977, Cambridge: Cambridge University Press.

Robinson, D., The Process of Becoming Ill, 1971, London: Routledge and Kegan Paul.

Sagarin, E., "Voluntary Associations Among Social Deviants", Criminolgia, 1967, 5, 8-22.

Sagarin, E., Odd Man In, 1969, Chicago: Quadrangle Books.

Saunders, C., ed. The Management of Terminal Malignant Disease, 1977, London: Edward Arnold.

Saunders, C., "Dying they Live: St Christopher's Hospice", in Feifel, H., New Meanings of Death, 1977, New York: McGraw Hill, 154-178.

Scheff, T., "Decision Rules and Types of Error, and their Consequences in Medical Diagnosis", Behavioural Science, 8, 1963, 97-107, in Freidson, 1972.

Schenthel, J.E., "Multiphasic Screening of the Well Patient", Journal of the American Medical Association, 172, 1960, 1-4.

Schoenberg, B., et al. Anticipatory Grief, 1974, New York: Columbia University Press.

Schoenberg, B., Carr, A.C., Peretz, D., Kutscher, A.H., eds. Psychological Aspects of Terminal Care, 1972, New York and London: Columbia University Press.

Seligman, J., (with Agrest, S.), "A Death in the Family", Newsweek, June 20: 89.

Seligman, M.E.P., "Submissive Death: Giving Up on Life", Psychology Today, 7, 1974, 80-85.

Sheskin, A., Why Die? Extend Your Life Through Cryonics: A Study of Belief Systems and Bereavement, 1979, New York: Irvington Publishers, Inc.

Stannard, D.E., "Death and Dying in Puritan New England", American Historical Review, December, 1973, 1305-1330.

Stannard, D.E., Death in America, 1975, Pennsylvania: University of Pennsylvania Press.

Stoddard, S., The Hospice Movement, 1978, New York: Vintage Books.

To my children
Cathy, Kerry and Peris Jones
and especially to my wife Pat

IDEOLOGICAL GROUPS